T0291804

CAMBRIDGE LIBRARY COLLECTION

Books of enduring scholarly value

Botany and Horticulture

Until the nineteenth century, the investigation of natural phenomena, plants and animals was considered either the preserve of elite scholars or a pastime for the leisured upper classes. As increasing academic rigour and systematisation was brought to the study of 'natural history', its subdisciplines were adopted into university curricula, and learned societies (such as the Royal Horticultural Society, founded in 1804) were established to support research in these areas. A related development was strong enthusiasm for exotic garden plants, which resulted in plant collecting expeditions to every corner of the globe, sometimes with tragic consequences. This series includes accounts of some of those expeditions, detailed reference works on the flora of different regions, and practical advice for amateur and professional gardeners.

Conversations on Vegetable Physiology

Jane Haldimand Marcet (1769–1858) wrote across a range of topics, from natural philosophy to political economy. Her educational books were especially intended for female students, to combat the prevalent idea that science and economics were unsuitable for women, but they found broader popularity: Michael Faraday, as a young bookbinder's apprentice, credited Marcet with introducing him to electrochemistry. This two-volume work, first published in 1829, is a beginner's guide to botany. Since the chief aim was accessibility, Marcet does not dwell on the often burdensome process of plant classification, but focuses on plant forms and botany's practical applications. She presents the facts in the form of simple conversations between two students and their teacher. Based on the lectures of the Swiss botanist Candolle, Volume 1 introduces roots, leaves, sap, and the effects of different soil and air.

Conversations on Vegetable Physiology

Comprehending the Elements of Botany, with their Application to Agriculture

VOLUME 1

JANE HALDIMAND MARCET

CAMBRIDGE
UNIVERSITY PRESS

University Printing House, Cambridge, CB2 8BS, United Kingdom

Published in the United States of America by Cambridge University Press, New York

Cambridge University Press is part of the University of Cambridge.
It furthers the University's mission by disseminating knowledge in the pursuit of education, learning and research at the highest international levels of excellence.

www.cambridge.org
Information on this title: www.cambridge.org/9781108067454

This edition first published 1829
This digitally printed version 2014

ISBN 978-1-108-06745-4 Paperback

Selected botanical reference works available in the CAMBRIDGE LIBRARY COLLECTION

al-Shirazi, Noureddeen Mohammed Abdullah (compiler), translated by Francis Gladwin: *Ulfáz Udwiyeh, or the Materia Medica* (1793) [ISBN 9781108056090]

Arber, Agnes: *Herbals: Their Origin and Evolution* (1938) [ISBN 9781108016711]

Arber, Agnes: *Monocotyledons* (1925) [ISBN 9781108013208]

Arber, Agnes: *The Gramineae* (1934) [ISBN 9781108017312]

Arber, Agnes: *Water Plants* (1920) [ISBN 9781108017329]

Bower, F.O.: *The Ferns (Filicales)* (3 vols., 1923–8) [ISBN 9781108013192]

Candolle, Augustin Pyramus de, and Sprengel, Kurt: *Elements of the Philosophy of Plants* (1821) [ISBN 9781108037464]

Cheeseman, Thomas Frederick: *Manual of the New Zealand Flora* (2 vols., 1906) [ISBN 9781108037525]

Cockayne, Leonard: *The Vegetation of New Zealand* (1928) [ISBN 9781108032384]

Cunningham, Robert O.: *Notes on the Natural History of the Strait of Magellan and West Coast of Patagonia* (1871) [ISBN 9781108041850]

Gwynne-Vaughan, Helen: *Fungi* (1922) [ISBN 9781108013215]

Henslow, John Stevens: *A Catalogue of British Plants Arranged According to the Natural System* (1829) [ISBN 9781108061728]

Henslow, John Stevens: *A Dictionary of Botanical Terms* (1856) [ISBN 9781108001311]

Henslow, John Stevens: *Flora of Suffolk* (1860) [ISBN 9781108055673]

Henslow, John Stevens: *The Principles of Descriptive and Physiological Botany* (1835) [ISBN 9781108001861]

Hogg, Robert: *The British Pomology* (1851) [ISBN 9781108039444]

Hooker, Joseph Dalton, and Thomson, Thomas: *Flora Indica* (1855) [ISBN 9781108037495]

Hooker, Joseph Dalton: *Handbook of the New Zealand Flora* (2 vols., 1864–7) [ISBN 9781108030410]

Hooker, William Jackson: *Icones Plantarum* (10 vols., 1837–54) [ISBN 9781108039314]

Hooker, William Jackson: *Kew Gardens* (1858) [ISBN 9781108065450]

Jussieu, Adrien de, edited by J.H. Wilson: *The Elements of Botany* (1849) [ISBN 9781108037310]

Lindley, John: *Flora Medica* (1838) [ISBN 9781108038454]

Müller, Ferdinand von, edited by William Woolls: *Plants of New South Wales* (1885) [ISBN 9781108021050]

Oliver, Daniel: *First Book of Indian Botany* (1869) [ISBN 9781108055628]

Pearson, H.H.W., edited by A.C. Seward: *Gnetales* (1929) [ISBN 9781108013987]

Perring, Franklyn Hugh et al.: *A Flora of Cambridgeshire* (1964) [ISBN 9781108002400]

Sachs, Julius, edited and translated by Alfred Bennett, assisted by W.T. Thiselton Dyer: *A Text-Book of Botany* (1875) [ISBN 9781108038324]

Seward, A.C.: *Fossil Plants* (4 vols., 1898–1919) [ISBN 9781108015998]

Tansley, A.G.: *Types of British Vegetation* (1911) [ISBN 9781108045063]

Traill, Catherine Parr Strickland, illustrated by Agnes FitzGibbon Chamberlin: *Studies of Plant Life in Canada* (1885) [ISBN 9781108033756]

Tristram, Henry Baker: *The Fauna and Flora of Palestine* (1884) [ISBN 9781108042048]

Vogel, Theodore, edited by William Jackson Hooker: *Niger Flora* (1849) [ISBN 9781108030380]

West, G.S.: *Algae* (1916) [ISBN 9781108013222]

Woods, Joseph: *The Tourist's Flora* (1850) [ISBN 9781108062466]

For a complete list of titles in the Cambridge Library Collection please visit:
http://www.cambridge.org/features/CambridgeLibraryCollection/books.htm

CONVERSATIONS

ON

VEGETABLE PHYSIOLOGY.

VOL. I.

CONVERSATIONS

ON

VEGETABLE PHYSIOLOGY;

COMPREHENDING

THE ELEMENTS OF BOTANY,

WITH

THEIR APPLICATION TO AGRICULTURE.

BY THE AUTHOR OF

"CONVERSATIONS ON CHEMISTRY," "NATURAL PHILOSOPHY,"

&c. &c.

IN TWO VOLUMES.

VOL. I.

LONDON:

PRINTED FOR

LONGMAN, REES, ORME, BROWN, AND GREEN,

PATERNOSTER-ROW.

1829.

LONDON:
Printed by A. & R. Spottiswoode,
New-Street-Square.

PREFACE.

THE favourable reception which the former works of the Author have met with, encourages her to offer these volumes to the public. In doing so, she cannot but feel diffident of success, as the subject of the present work is one, with which she has but recently become acquainted. Yet the source from which her knowledge of the vegetable creation is derived, makes her hope, that this new mode of studying Botany may be found interesting and useful.

The reader will perceive that the facts and opinions contained in the following pages, are almost exclusively taken from the lectures of a distinguished Professor of Geneva. To him, indeed, whatever merit may be found in this work, is due. The instruction and amusement which the Author derived from his lectures, led her to think, that she might, under the form of Conversation, convey a part of that interest to the minds of others. All she can lay claim to, is having arranged the subject in that form, which has always appeared to her, to possess great clearness and advantage, in fixing the attention of young people.

In acknowledging her obligations for the encouragement and assistance which that friend has so kindly given her, she must at the same time consider herself responsible for any errors, or inaccuracies which, either through inattention or want of knowledge, may have crept into the work.

TABLE OF CONTENTS.

VOL. I.

CONVERSATION I.

INTRODUCTION.

CONVERSATION II.

ON ROOTS.

CONVERSATION III.

ON STEMS.

CONVERSATION IV.

ON LEAVES.

CONVERSATION V.

ON SAP.

CONVERSATION VI.

ON CAMBIUM, AND THE PECULIAR JUICES OF PLANTS.

CONVERSATION VII.

ON THE ACTION OF LIGHT AND HEAT ON PLANTS.

CONVERSATION VIII.

ON THE NATURALISATION OF PLANTS.

CONVERSATION IX.

ON THE ACTION OF THE ATMOSPHERE ON PLANTS.

CONVERSATION X.

ON THE ACTION OF WATER ON PLANTS.

CONVERSATION XI.

ON THE ARTIFICIAL MODES OF WATERING PLANTS.

CONVERSATION XII.

ON THE ACTION OF SOIL ON PLANTS.

CONVERSATION XIII.

THE ACTION OF SOIL ON PLANTS CONTINUED.

CONVERSATION XIV.

THE ACTION OF SOIL ON PLANTS CONTINUED.

PLATE I.

THE COMMON PEA.

Pisum vulgaris; Leguminous family, Dicotyledon.

a, Stem.
b b, Stipula.
c c, Petiole, or common leaf stalk.
d d, Folioles of the compound leaf.
e e, Apex of the leaf.
f f, Tendril, terminating the petiole of the leaf.
g g, Peduncle or flower-stalk, springing from the axilla, and dividing it into two pedicels.
i i, Pedicels.
k k, Axilla of the leaf.
l, The flower.
m m, The calyx.
n n, The corolla.
o, The standard or superior petal.
p p, The two wings or lateral petals.
q, The carina, or two lower petals soldered together, seen interiorly.
r, The torus, or base of the flower.
s, The stamens, nine of which are half soldered together by their filaments.
s^1, The tenth stamen free.
t t, The anthers.
v u w, The pistil — *v,* the ovary; *u,* the style; *w,* the stigma, bearded.

- *x*, The fruit or pod, of which a portion has been removed in order to show the seeds.
- *y y*, The seeds attached to the upper suture of the pericarp.
- *z z*, Firmeules or ligatures attaching the seeds to the pericarp.
- y^1, A seed detached.
- y^2, The cicatrice.
- y^3, The seed split open, showing the embryo plant and the two cotyledons.
- *A*, The radicle.
- *B B*, The two fleshy cotyledons.
- *C*, The plumula.

PLATE II.

WILD TULIP.

Tulipa sylvestris; family Liliaceous, Monocotyledon.

a, Stem.
b, That part of the stem which forms the peduncle of the flower.
c c c, Leaves.
d d, Flowers with six pieces disposed in two rows, bearing the name of Perigone.
e, Torus, or base of the organs of the flower.
f, Filaments of the stamens bearded at their base.
f^1, The anthers.
g, The pistil composed of the ovary and the stigma, having no style.
g^1, The ovary.
g^2, The stigma crowning the ovary, composed of three cells.
C c, The pistil enlarged and grown into a fruit.
$C\,c^1$, The cut open, to show the three cells, separated by partitions, and enclosing each two rows of seeds attached to the centre of each cell.
h, A separate seed.
h^1, The same cut through lengthwise, to show the spermoderm, the albumen, and the embryon.
i, The spermoderm.
k, The albumen.

l, The embryon.
l[1], The embryon alone, showing it to be of one piece or monocotyledon.
m, The bulb.
n, The base, representing the trunk or stem.
o, The roots.
p, A lateral branch.

PLATE III.

CHINA-ASTER.

Aster Chinensis; Syngenesios family, or Compound Flower.

a, The stem.
b, A branch.
c c, Leaves.
d, A head not blown.
d^1, A head in blossom.
e e, Folioles composing the involucre.
f, Floral leaves, approximating to the form of the folioles of the involucre.
g, Ligulate florets situated around the disk.
g^1, A single floret remaining on the head, all the others being taken off.
h, Tubular florets situated on the disk or centre of the head.
h^1, A single tubular floret remaining on the disk.
g^{11}, A ligulate floret magnified.
i, Tube of the calyx soldered on the ovary.
k, Edges of the calyx terminating in layers or pappus.
l, Ligulate petal terminating in fine teeth.
m, Two stigmas.
h^{11}, A tubular floret magnified.
i^1, Tube of the calyx soldered on the ovary.
k^1, Pappus crowning the calyx.

l^{1}, Tubular petal terminating in fine teeth.

m^{1}, Stigma s.

n, Upper part of the style bearing the stigmas.

n^{1}, The same magnified.

n^{11}, The lower part of the style.

m^{11}, The two stigmas enlarged to see the sweeping hairs.

l^{11}, Tubular petal split lengthways and spread open.

o o, The fine filaments of the stamens.

p p, The five anthers soldered together, and forming a tube.

q, The fruit entire crowned by the pappus.

r, The fruit magnified.

s, The cicatrice, by which the fruit adheres to the receptacle.

t, The border of calyx magnified, showing a single hair truncated, inserted in a ring of teeth, the rest of the hairs being pulled off.

u, The embryon, in which may be distinguished the radicle and the two cotyledons.

x, Receptacle of the florets.

PLATE IV.

Fig. 1.

GERMINATION OF A MONOCOTYLEDON, OR ENDOGENOUS PLANT.

The *Scheuchzeria palustris.*

a, Pivot or radicle.
b b, Accessory roots shooting from the bottom of the stem.
c, Cotyledon or first leaf.
d d^1, Second and third leaves, called primordial.
c c, Common leaves of the plant.

Fig. 2.

HORIZONTAL SECTION OF THE STEM OF A MONOCOTYLEDON, OR ENDOGENOUS PLANT.

Yucca aloifolia.

Showing the scattered fibres which compose the wood, having neither bark, pith, medullary rays, nor distinct layers.

Fig. 3.

GERMINATION OF A DICOTYLEDON, OR EXOGENOUS PLANT.

Daubnentonia punic a.

a, Radicle slightly branching.
b, Neck or vital point between the root and the stem.
t, Portion of the stem below the cotyledons.
T, Portion of the stem above the cotyledons
c c, Two opposite cotyledons.
d, A simple primordial leaf.
ff, Common leaves.

Fig. 4.

VERTICAL SECTION OF THE STEM OF A DICOTYLEDON, OR EXOGENOUS PLANT.

The *Oak.*

a b, The bark, composed of the vertical layers *a,* and the internal bark *b.*
c d e, The wood, composed of the alburnum or young wood *c,* the perfect wood *d,* and the pith *e.*
The circular zones represent the layers of wood, and the lines diverging from the centre the medullary rays.

Fig. 5.

A BRANCH TURNING ITS LEAVES TOWARDS THE LIGHT.

CONVERSATIONS.

CONVERSATION I.

EMILY.

As I wander over these beautiful mountains, and observe the variety of flowers they produce, how much I regret my ignorance of botany!

MRS. B.

It is, certainly, a science particularly adapted to Switzerland; but why should you suffer your regret to be vain? To wish to learn is the first, and often the most difficult step towards the acquisition of knowledge.

EMILY.

I should certainly like to understand botany, but I have no wish to learn it: there is such a drudgery of classification to encounter, before one can attain any proficiency to recompense one's labours, that I confess I do not feel courage to make the attempt.

CAROLINE.

And, after all, what is it one acquires? — A knowledge of the class in which a flower is to be placed, according to the number of its stamens or its pistils; and, perhaps, after hard study, one may go so far as to ascertain its Latin name, though you may still be ignorant how it is called in your own vulgar tongue. Botany appears to me a science of rules and names, not of ideas; and is, therefore, devoid of interest. I am, for my part, quite contented to gather a sweet-smelling nosegay of beautiful garden monsters, as botanists denominate them, without troubling myself about their scientific names.

MRS. B.

I will frankly own, that, for many years, I entertained the same prejudices against botany, if such you will allow me to call them; but having had the good fortune, during a spring I passed at Geneva, to hear a course of lectures on that science by Professor De Candolle, I was entirely converted; and I am fully persuaded that no natural science is dry, unless it be dryly treated. If people will attend more to the frame than to the picture which it contains, and if they will even cut and disfigure the picture, in order to make it fit into the frame they have prepared for it, no wonder that the subject should lose its interest.

EMILY.

None can be more likely to succeed in converting others, than those who have been converted themselves; and if you would indulge us, my dear Mrs. B., with relating what you learnt at these lectures, I make no doubt that Caroline would be tempted to listen to you, were it but from curiosity to discover whether her first opinions on the subject were correct, or whether she ought not, at least, to acknowledge that they were hastily formed.

CAROLINE.

Oh, I shall be very thankful to be allowed to remain, provided I am at liberty to depart if I find I do not take an interest in the study.

MRS. B.

I shall not be ambitious of retaining uninterested listeners; and though I was delighted with the instruction I received myself, I am very sensible that I shall not be able to communicate to you either the same degree of pleasure or of information. I will, however, do my best to relate to you what I can recollect of these lectures.

Mr. De Candolle, so far from confining himself to the classification of plants, examines the vegetable kingdom in its most comprehensive and philosophic point of view. In describing the structure he investigates the habits and properties of plants, and shows, not only how wonderfully they have been formed to answer the purposes of their own multiplication and preservation, but how admirably they answer the higher purpose which Nature has assigned them, of ministering to the welfare of a superior order of beings — the animal creation; and more especially to that of man. He turns his attention particularly to point out the means by which the science of botany can promote that with which it is most intimately and importantly connected — agriculture. He prepares the soil and sows the seed for the husbandman; he extracts the healing juices and the salutary poisons for the physician; he prepares materials for the weaver, colours for the dyer; in a word, as he proceeds, there is scarcely an art on which he does not confer some benefit, either by pointing out a new truth, or warning against an ancient error. Thus, throughout his course, his principal aim is to promote, by his vast stock of knowledge, the welfare of his fellow-creatures.

EMILY.

Treated in this point of view, botany cannot, I think, fail to interest us.

MRS. B.

It is rather the physiology of botany which I propose teaching you, and I shall merely give you such an insight into classification as is necessary to enable you to understand the structure and character of plants.

Mr. De Candolle entered upon the subject by observing, that, in classing the various productions of Nature, the first great line of demarcation is that which separates the mineral kingdom from organised beings. How would you make the distinction?

CAROLINE.

Nothing more obvious: organised beings have life, and minerals have not.

MRS. B.

Very true; yet I should be tempted to retort upon you that this distinction is rather of names than of ideas. I believe I have before observed to you, that we know not what life is. We see its effects: they are sufficiently apparent and numerous; and it is only by studying these effects that we are able to form any idea of that state of being which we call life. The first distinction, therefore, to be made between minerals and beings endowed with life is, that the latter are formed with organs adapted to fulfil the several functions for which they were destined by Nature. These organs differ, not only in form and structure, but more or less in the materials of which they are composed: organised beings are generally of a smooth surface, rounded, and irregular; whilst minerals are rough, angular, and in their crystalline state of geometrical regularity.

One of the principal functions these organs have to perform is nutrition. Unorganised matter may, in the course of nature, be enlarged or diminished, either by mechanical or chemical changes; minerals may be augmented by the addition of similar particles, or by chemical combination with substances which are dissimilar, but they have no power to convert them into their own nature.

Organised bodies, on the contrary, are increased in size, by receiving internally particles of matter of a nature different from their own, which they assimilate to their own substance.

EMILY.

That is to say, that the food by which they are nourished is converted into their own substance?

MRS. B.

Yes; organised beings have also the power of reproducing their species: — minerals may be broken into fragments, but they are alike incapable of receiving nourishment, of growing, or of reproducing.

Let us now proceed to enquire, what is the principal distinction between the two classes of organised beings, the animal and the vegetable creation.

CAROLINE.

Animals are endowed with a power of locomotion, whilst vegetables are attached to the soil.

MRS. B.

It would, perhaps, be more philosophical to begin

by ascertaining the cause whence this difference arises. Animals are provided with a cavity called a stomach, in which they deposit a store of food, whence they are continually deriving nourishment. This organ is essential to animals, as they are not constantly supplied with food: they find it not always beneath their feet; they must wander in search of it; and were they not furnished with such a storehouse, in which to lay it up, they would be frequently in danger of perishing.

EMILY.

Are we, then, in want of continual nourishment? And should we die if our stomachs were quite empty?

MRS. B.

No, not immediately; for though the system requires constant renovation, Nature is so careful of our preservation, that she not only affords us the means of subsistence, but provides resources in cases of accidental interruption of the supply: after having consumed, or rather, I should say, assimilated the food contained in the stomach, the fat of animals is made to contribute to the nourishment of their organs, and the support of life. In some, such as the dormouse and the polar bear, this provision is carried to such an extent, that they pass several of the winter months in a state of inanition; during which period, the only sustenance their system receives is from the abundant provision of fat which they had made during the summer; and when they are roused from their lethargy by the return of spring, they are lean and ravenous.

The food of animals is conveyed from the stomach to the various parts of the body by the function which is called *digestion*. The food passes through small absorbent vessels into the blood, and is thence circulated throughout the system.

CAROLINE.

But, Mrs. B., one would think you were going to give us the history of the animal rather than the vegetable creation.

MRS. B.

Only so far as to enable me to point out the distinction between them.

Vegetables have no stomach; they do not require such a magazine, since they find a regular supply of nourishment at the extremity of their roots: with them, therefore, there is no occasion for accumulation. In order to conceive an idea of the form in which plants receive nourishment, you must represent to yourself a very delicate cobweb network, of such extreme tenuity as to render it invisible until the interstices are filled and distended by the nutriment lodged within them. The food of plants is not like that of animals, of a complicated nature; but consists of the simplest materials, — water, and the solid and gaseous matter contained within it.

The second distinction between the animal and vegetable creation is, that the latter are not endowed with sensibility.

EMILY.

But the mimosa or sensitive plant, Mrs. B., when it shrinks from the touch, wears a strong appearance of sensibility.

MRS. B.

Yet I should doubt whether it is any thing more than appearance. Some ingenious experiments have, indeed, been recently made, which tend to favour the opinion that plants may be endowed with a species of sensibility; and seem to render it not improbable that there may exist in plants something corresponding with the nervous system in animals.

CAROLINE.

The sensitive plant would then, no doubt, be a nervous fine lady at the court of Flora. But, pray, of what nature were these experiments?

MRS. B.

There are certain vegetable poisons, such as nux vomica, laurel-water, belladonna, hemlock, and several others, which are known to destroy life in animals, not by affecting the stomach, but merely by acting on the nervous system. These poisons were severally administered to different plants, either by watering them with or steeping their roots in infusions of these poisons. The universal effect was, to produce a sort of spasmodic action in the leaves, which either shrunk or curled themselves up; and, after exhibiting various symptoms of irritability during a short time, became flaccid, and the plant in the course of a few hours died.

EMILY.

I should have been curious to have seen an experiment of this nature tried on the sensitive plant.

MRS. B.

It was done. Two or three drops of prussic acid, which, you know, is a most powerful poison, were poured upon a sensitive plant: the leaflets closed and opened again at the end of a quarter of an hour; but they did not regain their sensitiveness for at least six or eight hours. When we see plants thus acted upon by vegetable poisons, which are known to be incapable of destroying the animal fibre, or of injuring the frame but through the medium of the nerves, we may be led to suppose, that certain organs may exist in plants with which we are totally unacquainted, and which bear some analogy to the nervous system in animals.

It is certain that plants possess a power of irritability or contractibility; for it is by alternate contractions and dilatations of the vessels that they propel the juices which rise within them. Here is a slip of elder: when I cut it in two, the fluid continues oozing from both of the separated parts; were there no action going on within the stem, only a single drop would flow out at each orifice. There are some flowers, such as those of the Barberry, whose stamens will bend and fold over the pistil, if the latter be pricked with a needle; and there is one instance of a plant whose leaves move without any assignable cause: this is the *Hedysarum gyrans*, which grows only on the banks of the Ganges; it has three leaflets on each foot-stalk, all of which are in constant irregular motion.

EMILY.

I recollect seeing a plant called Sundew (*Drosera*), the leaves of which, near the root, are covered with bristles bedewed with a sticky juice. If a fly settles on the upper surface of the leaf, it is at first detained by this clammy liquid, and then the leaf closes, and holds it fast till it dies.

MRS. B.

The *Dionæa muscipula* affords another example of the same kind: it grows in the marshes of South Carolina. Its irritability is so great, that an insect which settles on it is generally crushed to death by the collapsing of the two sides of the leaf, which, like that of the Drosera, is armed with bristles.

CAROLINE.

But all plants are endued with some degree of irritability, if you will not admit of sensibility; for we know that, in general, their leaves turn towards the light, and, when growing in a room, they spread out their branches towards the windows, as if they were sensible of the benefits they derived from light and air.

MRS. B.

Light and air conduce to their wellbeing, and they are so wisely constructed by Providence as to seek them; but it is independently of all choice or preference. We must consider plants as beings in which the principle of life is reduced to its state of greatest simplicity. As we advance in the scale of creation, we find that the lowest animals are directed by instinct; intelligence increases as we approach towards man, who is guided by reason: but the vegetable world is influenced merely by physical causes, which derive their energy from the principle of life.

EMILY.

But since plants are so inferior in the scale of existence, why is their form so much more delicate and beautifully varied than that of animals? Is it not singular that Nature should be most solicitous for the appearance of her simplest works?

MRS. B.

The most curious details of the structure of a plant are visible in its outward form; whilst those of the animal economy are concealed in the anatomical structure of the internal parts. The organs of plants are chiefly external, and are ornamental at the same time that they perform the several functions for which they were formed.

Plants appear, also, to be susceptible of contracting habits: the mimosa, or sensitive plant, if conveyed in a carriage, closes its leaves as soon as the carriage is in motion, but after some time it becomes accustomed to it, the contraction ceases, and the leaves expand; but if the carriage stops for any length of time, and afterwards recommences its motion, the plant again folds its leaves, and it is time only which can reconcile it to its new situation.

EMILY.

This evinces strong symptoms of sensibility. One would suppose that the plant was alarmed at the new and unknown state of motion; and that its apprehension, like that of an infant, returned every time the novelty recurred.

MRS. B.

You will, perhaps, consider plants as patriotic, when you learn that those which are brought from the southern hemisphere, faithful to the seasons of their native country, make vain attempts to bud and blossom during our frosty winter, and seem to expect their sultry summer at Christmas.

CAROLINE.

If you continue thus, Mrs. B., you will certainly make me think that plants are not wholly devoid of sensibility.

MRS. B.

We cannot positively deny it; but the evidence against that opinion is so strong as to amount almost to proof. Had Providence endowed plants with the sensations of pleasure and of pain, he would, at the same time, have afforded them the means of seeking the one and of avoiding the other. Instinct is given to animals for that express purpose, and reason to man; but a plant rooted in the earth is a poor, patient, passive being: its habits, its irritability, and its contractibility, all depending on mere physical causes.

The properties of plants may be separated into two classes: first, those which relate to their structure; such as their elasticity, their hygrometric power: these properties may continue after death. Secondly, those which relate to their vitality; such as contractibility: which, consequently, can exist only in the living state.

The organs of vegetables are all composed of a membranous tissue, which pervades the whole of the plant; they are distinguished by the name of elementary, and are of three kinds.

1st. The cellular system, consisting of a fine tissue of minute cells or vesicles, of an hexagonal form, apparently closed and separated by thin partitions, somewhat similar to the construction of a honeycomb; or bearing, perhaps, a still nearer resemblance to the bubbles formed by the froth of beer.

EMILY.

This appears very similar to the cellular system in the animal economy, which you described to us in our lessons on Chemistry.

MRS. B.

One of the chief purposes of the cellular system in the animal frame is to contain the fat, a substance to which there is nothing analogous in the vegetable kingdom.

These cells in plants are marked by small spots, which have been conjectured to be apertures through which fluids are transmitted from one cell to another; but these marks are so very minute, as to render it hazardous to venture on deciding for what purpose they are designed.

CAROLINE.

If it is the cellular system which transmits the sap, it should with more propriety be compared to the veins and arteries of animals. But are not plants furnished also with a vascular system?

MRS. B.

Yes; and this forms the second set of elementary organs. It consists of tubes open at both ends: these are always situated internally, and are, besides, guarded from injury by being lodged in a thick coating of the cellular integument. Some of these vessels assume the form of a necklace, their coats being at intervals drawn tight together, or strangulated, so as to appear to stop the passage of the fluid they contain.

CAROLINE.

It is doubtless through the vascular system that the sap rises?

MRS. B.

The organs of plants are so extremely small, that, though aided by the most powerful microscope, it is frequently difficult to examine the structure of their parts with a sufficient degree of accuracy to be able to ascertain their functions. It has long been a disputed point, whether the sap ascended through the vascular or the cellular system of organs; but the latest opinion, and that which Professor De Candolle is inclined to favour, is, that it passes through neither; and that it rises through interstices which separate the different cells.

EMILY.

Indeed! It seems to me very extraordinary that the sap, which performs so essential a part in the economy of vegetation, should not flow freely through appropriate vessels, but be left to find its way as it can between them.

MRS. B.

The sap, when first pumped up by the roots, consists of little more than water, holding various crude materials in solution; it is, therefore, more important that the regular organs should be reserved for its elaboration, and its conveyance after that process, to the several parts of the plant.

The third system of elementary organs is the tracheæ; so called from their conveying air both to and from the plant: they are composed of very minute elastic spiral tubes.

CAROLINE.

But, surely, plants do not breathe, Mrs. B.?

MRS. B.

Not precisely in the same manner that we do; but air is so essential an agent, both chemically and mechanically, in promoting their nourishment and growth, that it is scarcely less necessary to their existence than to that of animals. Indeed, it is the opinion of Professor De Candolle, that the function of transmitting air is not confined to the tracheæ, but extends to the whole of the vascular system.

The whole of the vegetable kingdom consists of masses of these several elementary organs, with the exception of fungi, mosses, and lichens, whose vessels are all of a cellular form: they have no vascular system whatever.

EMILY.

That affords a strong argument against the passage of the sap through the vascular system.

MRS. B.

Certainly; the fibres of plants are composed of collections of these vessels and cells closely connected together. The root and stem of plants consist of such fibres: if you attempt to cut them transversely, you meet with considerable resistance, as you must force your way across the tubes, and break them; whilst, if you slit the wood longitudinally, you separate the vessels without breaking them, and have only to force your way through the elongated cellular tissue which connects them.

EMILY.

The difference is very observable; but I wonder that the cells, being formed of a delicate membrane, are not squeezed and crushed to pieces in the stems of plants, especially when they become hard wood.

MRS. B.

The cells, by the growth of the stem, are frequently drawn out of their original form, and elongated; but the vascular system, which is of the greatest importance, is internal, and lodged in a bed of cellular integument, so that the pressure of the bark or surrounding parts is not sufficient to crush it.

The layers of wood which you may have noticed in the stem or branch of a tree cut transversely, consist of different zones of fibres, each the produce of one year's growth, and separated by a coat of elongated cellular tissue, without which you could not well distinguish them.

The cuticle, which is the external skin or covering of the plant, consists of an expansion of the cellular tissue; and is furnished with pores for evaporation.

CAROLINE.

This is, I suppose, neither more nor less than what is commonly called the bark?

MRS. B.

On the contrary, it is both *more* and *less* than the bark. *More*, because the cuticle is extended over every part of the plant; it covers the leaves and flowers, with the exception of the pistil and anthers, as well as the stem and branches; *less*, because the bark consists of three distincts coats, of which the cuticle forms only that which is external. The cuticle of a young shoot, after it has been for some time exposed to the atmosphere, becomes opaque, dries, and distended by the lateral growth of the branch, splits, and after a year or two falls off. A second membrane is then formed by the desiccation of the external part of the cellular integument; but it differs from the former in being thicker, and of a closer texture. It is not furnished with pores, having no other function to perform than to enclose a layer of air, and preserve the internal parts from injury. This envelope is distinguished from the former by the name of epidermis.

These general, though, perhaps, rather desultory observations will, I hope, prepare you for our next interview; when I propose to take a full-grown plant, examine its structure, and explain the nature of those organs by which it is nourished and preserved. We shall begin with the roots, and then proceed up the stem to the leaves.

EMILY.

I should have expected that you would have

commenced by the birth of the plant, that is to say, the germination of the seed.

MRS. B.

If the plant derives life from the seed, the seed equally owes its origin to the parent plant; and as the preparation of the seed, by that beautiful and delicate system of organs, the flower, is one of the most curious and complicated operations of the vegetable economy, I think it more eligible to reserve it for the latter part of our studies.

CAROLINE.

That is very true so far as regards the formation of the seed; but its bursting, and the sprouting of the young plant, appears to be the natural commencement of the history of vegetation.

MRS. B.

The germination of the seed is a process so intimately connected with its formation and composition, that it is a reciprocal advantage to treat of them together, or, rather, in immediate succession, instead of separating them by the intervention of the whole history of vegetation.

CONVERSATION II.

ON ROOTS.

MRS. B.

We are now to examine the structure of those organs, whose office it is to nourish and preserve the plant.

In the nutrition of plants, six periods are to be distinguished: —

1. The absorption of nourishment by the roots.
2. The transmission of nourishment from the roots to the different parts of the plant.
3. The developement of the nourishment.
4. The action of the air on plants.
5. The conversion of nourishment into returning sap or cambium.
6. The secretion of various juices from the sap.

Plants being deprived of locomotion, as we have observed, cannot go in search of food: it is necessary, therefore, that nature should provide it for them in their immediate vicinity. Those simple elements, which are almost every where to be met with, water and air, constitute this food. Water not only forms the principal part of it, but serves, also, as a vehicle to convey what solid food the plant requires; and as a vegetable is unfurnished either with a mouth to masticate, or a stomach to digest, solid food can be received only when dissolved in water. In this state it is absorbed by the roots; for the root not only supports the plant by fixing it in the soil, but affords a channel for the conveyance of nourishment. If it does not fulfil this double office, it is not a root, but a subterraneous branch.

CAROLINE.

But will not a branch, if placed under ground, become a root, and absorb nourishment? I have seen the gardener fasten down branches of laurel and other shrubs, leaving only the extremity above ground; and these layers strike root, and become, in the course of time, separate plants.

MRS. B.

Striking root implies, that roots will (under certain circumstances) grow from a branch, but the branch itself cannot be converted into a root; for at the extremity of each fibre of a root, there is an expansion of the cellular integument called a spongiole, from its resemblance to a small sponge, being full of pores, by means of which the roots absorb the water from the soil. Now, a branch, being destitute of this apparatus, cannot supply the plant with nourishment.

CAROLINE.

True: it cannot feed without a mouth;—but I thought that there were pores in every part of a plant.

MRS. B.

The pores in those parts of a plant above ground are almost wholly for the purpose of exhalation. The roots have no pores except in the spongioles at their extremities, which, as I have observed, are for the purpose of absorption. It would be very useless for them to be furnished with evaporating pores, since they are not exposed to the atmosphere, where alone evaporation could take place.

EMILY.

The tendrils of vines, then, and of other climbing plants, which serve to fix them against a wall, or the trunk of a tree, cannot be considered as roots; since, although they answer the purpose of sustaining the plant, they are unable to supply it with nourishment.

MRS. B.

Certainly, these plants are furnished with roots which pump up nourishment from the soil; but there are some parasitical plants, such as the *Viscum album* or misletoe, and the *Epidendron Vanilla*, which, having no immediate communication with the earth, strike their fibres into the stems or branches of a tree, and derive their nourishment from this richly-prepared soil; but as the absorption in this case is not carried on by the regular mode of spongioles, these fibres are not denominated roots.

A root is never green, even when exposed to the light, an element which is essential to the developement of the green colour in other parts of the plant.

The root, then, by means of the little spongioles attached to its extremities, sucks up whatever liquid comes within its reach; in proportion as it grows, its fibres spread themselves over a greater extent of soil, and come in contact with a greater quantity of moisture; and the plant, whose branches extend above ground, in proportion as the root spreads beneath, requires a more abundant supply of food.

EMILY.

And do the roots take up every kind of liquid, or have they any means of selecting what is suited for their nourishment?

CAROLINE.

How would it be possible for them to make a choice, having neither reason nor instinct to direct them? For I conclude that the little spongioles are not endowed with the sense of taste, to enable them to discriminate between different sorts of food.

EMILY.

True; but without endowing the vegetable creation with reason, instinct, or even sensibility, Nature might possibly have so constructed the absorbent pores, that, either by mechanical or chemical means they should reject what was unfit, and receive only what was good for the plant.

MRS. B.

The only provision which Nature appears to have made with this view, is, to have formed the pores of the spongioles of such small dimensions, that they are incapable of absorbing a liquid which is thick or glutinous; for if the fluid be loaded with particles not extremely minute, they cannot pass through the tubes which compose the vascular system of the plant. I do not mean to say that these pores have any power to reject a dense or viscous fluid, but that they will be clogged and obstructed by it, and the absorption consequently cease.

Water which has flowed through the manure of a farmyard, and abounds with nutritive particles, is much used on the Continent for watering gardens; yet, unless copiously diluted with pure water, it is found to be deleterious, choking the plant with an excess of food. But when the liquid is sufficiently limpid, the spongioles suck it up with equal avidity, whether it contain salubrious nourishment or deadly poison.

EMILY.

Oh, my poor plants! Why did not Nature grant them some means of preservation from such dangers?

MRS. B.

Nature has bountifully diffused throughout the soil such fluids as are adapted for the nourishment of the vegetable creation: no streams of poison flow within their reach. It is unnecessary, therefore, to guard against a danger which does not exist. It is merely from the experiments of the chemist and the physiologist that we learn that the roots of plants will absorb liquids, of whatever nature, presented to them, provided they be sufficiently limpid. The spongioles act only by capillary attraction, and suck up moisture just as a lump of sugar absorbs the water into which it is dipped. As a proof of this it has been shown, that if roots, saturated with moisture, be transplanted into very dry earth, the latter will absorb the moisture from the roots.

EMILY.

If so, why do not the roots continue to absorb moisture when the plants are dead, as well as when they are living. A sponge, or a lump of sugar, have no vital principle to stimulate them to draw up liquids?

MRS. B.

Neither does absorption immediately cease upon the death of a plant, as the blood ceases to circulate upon the expiration of animal life; but when the vessels through which the fluid should pass have lost their vital energy, that susceptibility of irritation and of contraction, which enabled them to propel the fluid upwards, ceases, and it can no longer ascend into the roots, but remains stagnant in the spongioles, which soon become saturated. Disease and putrefaction follow; and that nourishment, which was designed to sustain life, now serves only to accelerate disorganisation. The fluid is, however, still performing the part assigned to it by Nature; for if it be necessary to supply living plants with food, it is also necessary to destroy those which have ceased to live, in order that the earth may not be encumbered with bodies become useless, and that their disorganised particles may contribute to the growth of living plants. Thus the putrefaction of dead leaves, straw, &c. which reduces these bodies to their simple elements, prepares them to become once more component parts of living plants.

CAROLINE.

What a beautiful provision for the vegetable economy! I know not whether you call this botany, Mrs. B.; but it is totally different from the dry classification of flowers. It elevates the heart while it enlightens the mind, and bears more resemblance to lessons of morality and religion than to botany.

MRS. B.

The physiology of plants, of which we are now treating, forms one branch of the science of botany, and one which is certainly replete with interest; but from every natural science, and every branch of it, from the arrangement and classification of the organs of the flower as well as from the history of vegetation, the well-disposed mind will draw lessons of piety; and he must study Nature with very contracted views, who does not raise his thoughts from the admiration of the creation to that of its all-wise and beneficent Creator. But to return to our subject.

Botanists distinguish several kinds of roots. The *Radix fibrosa*, or fibrous root, is the most common and most simple in its form: it consists of a collection or bundle of fibres, connected by a common head, and often merely by the base of the stem. The roots of many grasses and most annual herbs are of this description; during their short existence, which is limited to one summer, they continue growing, both by forming new fibres and by elongating the old ones. These fibres are occasionally covered with a sort of shaggy down, which, as it generally occurs in loose or sandy soils, is considered as a provision of Nature for the purpose of fixing the plant more firmly in the ground.

EMILY.

Of what description are the roots of those weeds, such as couch-grass, which seem to be interminable. If you attempt to eradicate them, you meet with a succession of bunches of fibres springing from a root which grows horizontally, and appears to be endless.

MRS. B.

This is the *radix repens*, or creeping root. The long horizontal fibre is, in fact, not a root but a subterraneous branch, for it has no spongioles: the real roots are the small bunches of fibres which spring from it. Such a root is very tenacious of life, as any portion in which there is an articulation will grow: it decays at its origin, and continues growing at its extremity.

EMILY.

Then we must not seek for its origin but its extremity, in order to eradicate it.

MRS. B.

You cannot destroy it without digging up the whole of the subterranean stem: it is this which renders it so difficult to eradicate.

CAROLINE.

Yet surely not more difficult than the Ox-eye and many other weeds, whose strong penetrating roots seem to strike to the very centre of the earth; for, however loosened by digging, they are scarcely ever pulled up entire.

MRS. B.

The root of these plants is called *fusiform*, or spindle-shaped. It is also called the tap root, from its tapering so considerably towards the end; and the pivot-root, owing to its fixing the plant so immovably in the earth. This root is but scantily provided with the means of acquiring food, having sometimes not more than a single fibre furnished with a spongiole at its extremity. To compensate for this disadvantage, the root is of so moist and fleshy a nature as to afford an ample store of provision.

CAROLINE.

But with such limited means of suction, how can this magazine be replenished?

MRS. B.

The surface of the ground immediately exposed to the drying powers of the sun and wind, retains less moisture than the deeper and more sheltered strata of the soil; besides, the store is laid up during the season of abundance, and measured out, as the necessities of the plant require, during that of dearth. Here, you see, are a variety of compensations for its circumscribed power of absorption.

A very simple experiment will convince you, that the spindle-shaped root, as well as those of every other description, absorb water only by the spongioles at their extremities. If you immerse a young radish in a glass of water, so that every part of it shall be covered except the taper end of the root, you will find that it will soon die; while, if you immerse only the extremity of another radish in water, you will preserve it alive. The whole body of the root serves to fix and support the plant in the soil, but it is the extremity alone which absorbs nourishment.

It sometimes happens that this species of root, whether from want of vigour or some mechanical impediment, is checked in its growth, and wears the appearance of being cut or bitten off. It has hence obtained the name of *radix præmorsa*, or abrupt root; but it is, in fact, nothing more than the *radix fusiformis* originally mutilated, and modified by that mutilation in successive generations.

EMILY.

Is not the Devil's bit Scabious of this description? I recollect hearing a curious story of its acquiring this mutilated form. In the age of sorcery and credulity it was affirmed, that the devil, out of spite to mankind, bit off the end of a plant which was endowed with so many excellent properties.

MRS. B.

The name of the plant is, no doubt, derived from this ridiculous story; but I should be rather inclined to suppose that it was an allegorical compliment to the virtues of the plant, than that such absurdity could obtain belief in any age.

EMILY.

Bulbous roots, such as those of the hyacinth, the lily, and the onion, are also solitary roots, Mrs. B.; but they seem to fix the plant in the soil rather from their mass than their depth, for they are very superficial; and it is no doubt from the difficulty of finding water, that Nature has added to their root a tuft of small stringy fibres (which are doubtless furnished with spongioles) to multiply the points of absorption.

MRS. B.

The bulbous root, *radix bulbosa*, is improperly so called, for the tufts or fibres pendant from the bulb are alone the roots. The bulb itself, you will learn, when you come to examine its structure, constitutes the stem of the plant; no wonder, therefore, that it is superficial.

EMILY.

How curious! a globular subterraneous stem?

MRS. B.

If you prefer giving it the name of bud rather than of stem, you may with equal propriety, for it contains the whole embryo plant; but, as we are not at present treating the subject either of stems or of buds, we must reserve this explanation for a more appropriate period.

CAROLINE.

And are the roots of potatoes of this description?

MRS. B.

The potatoe belongs to the class of tuberous or knotted roots, *radix tuberosa*, which are of various kinds, comprehending all such as have fleshy knobs or tumours. This sort of root belongs to perennial plants, though the knobs are frequently either annual or biennial. In all cases, they are to be considered as reservoirs of nourishment, which enable the plant to support the casual privations of a barren or dry soil.

Some plants, of which Timothy grass is an instance, acquire tumours when situated in a soil subject to vicissitudes of draught and humidity, and lose them if transplanted to one regularly supplied with moisture.

EMILY.

It is wonderful to observe in what an admirable manner roots find means of compensation for local inconveniences!

MRS. B.

The potatoe is a species of mucilaginous, farinaceous excrescence, growing upon subterraneous branches, which have no means of deriving nourishment from the soil; and it is very remarkable, that this salubrious and nutritious substance grows on a plant, the real fruit of which is of a poisonous nature.

The object of nature, throughout all these varying forms, is the same — to establish a reservoir, in which the vital force of the plant and its material resources are husbanded.

The root of the *orchis* is well deserving our notice from its singularity. It consists of two lobes, somewhat similar to the two parts into which a bean is divided. One of these perishes every year, and another shoots up on the opposite side of the remaining lobe. The stem rises every spring from between the two lobes, and, since the new lobe does not occupy the same place as its predecessor, the orchis every year moves onwards, though to the distance only of a few lines.

CAROLINE.

Thus, in the course of a certain number of years, the orchis may make the tour of a garden, provided the gardener does not interrupt it in its progress.

MRS. B.

There are some plants which, like the Indian fig-tree, shoot out roots from the stem many feet above ground: they grow downwards, bury themselves in the soil, and new stems ultimately spring up from them; but the epidermis of these roots are never green, like that of young branches.

EMILY.

I recollect reading an account of a tree which bears some analogy to this fig-tree. It was situated at the top of a high wall, and its roots grew down the side of the wall till they reached the ground, a distance of about ten feet, and then buried themselves in the soil.

MRS. B.

This account is given by Lord Kaimes of a plane-tree, situated among the ruins of the New Abbey monastery in Galloway. But the analogy with the fig-tree is only apparent, this singular growth of the roots being merely the result of local inconvenience.

CAROLINE.

I once heard of a curious experiment performed on a willow-tree. It was dug up, and reversed, the head of the tree was planted in the ground, and the roots, which were now uppermost, stretched out like naked branches in the air. In the course of time, the roots were transformed into branches, and the branches into roots. But how could the latter acquire spongioles?

MRS. B.

They did not; but roots sprouted from the subterranean branches, and branches shot from the unburied roots. This is, however, an adventurous experiment, which does not often succeed.

The duration of roots is either annual, biennial, or perennial. To the first belong plants whose existence is limited to one summer, such as barley, and a vast number of garden and field flowers. The biennial root produces the first season only herbage, and the following summer flowers and fruit, or seed; after which it perishes. The perennial belong to plants which live to an indefinite period, such as trees and shrubs.

A root consists of a collection of fibres composed of vascular and cellular tissue, but without tracheæ or vessels destined for the transmission of air; but there is so great an analogy between the structure of the root and that of the stem, that I shall reserve what observations I have to make on this subject till our next meeting, when I propose to examine the nature of the stem.

CONVERSATION III.

ON STEMS.

MRS. B.

EVERY plant has a stem.

CAROLINE.

That is to say, trees and shrubs; for there are many plants, such as violets, anemones, fern, and a variety of others, which have large bunches of leaves growing from the roots out of the ground: the flowers, it is true, have each a stem, but the plant itself seems to have none.

MRS. B.

I must repeat my assertion: — every plant has a stem, through which the sap circulates, and from which the leaves and flowers spring. This stem, it is true, is not always apparent: it is sometimes concealed under ground, sometimes disguised under an extraordinary form. The stem of the tulip is contained within the bulb or onion, which is commonly, but improperly, called its root; that of the fern is subterraneous. A very curious plant grows in some of the vallies of the Alps, called willow-grass (*saule en herbe*). You sometimes meet with a plain covered with it, and you would not imagine whence it derives its origin: it is nothing less than the head, or rather, I should say, the extremities, of the branches of a large willow-tree.

EMILY.

Do you mean a tree which has been accidentally overthrown and buried, the leaves of which have sprouted above ground?

MRS. B.

No; it is a willow which is annually buried alive. Every spring it struggles to rise above ground, and every autumn it disappears beneath the soil. Let us suppose the seed of a willow springing up at the foot of a mountain, and that the earth which is annually carried down by the rains from this mountain should be sufficient to bury the young plant. The following spring it would again shoot out with redoubled vigour; for the growth of the plant having been checked by the fall of the soil, the sap, which should have been expended in the produce of foliage, being accumulated in the little stem, will be sufficient to afford nourishment for a double shoot; two little branches will therefore now appear. This, like its predecessor, flourishes but for a season, and is buried. The two stems the following spring produce four, which expand their leaves, and in the autumn are consigned to the earth; the third year eight stems arise; the fourth, sixteen; and the plant goes on thus doubling its sprouts every year, and the surface of the soil rising, till at length a plain is formed covered with verdure, consisting of the leaves of the willow-tree.

CAROLINE.

What a singular growth!—How much I should like to walk on one of these curious meadows!

MRS. B.

They are, as you may suppose, not very common, since it requires peculiar local circumstances to produce one: the vicinity of a mountain which shall annually send down earth sufficient to bury the young shoots, but not so deeply as to prevent their rising from their tomb every spring. The age of these willows has been ascertained by digging down the side of the plain and observing how often the shoots have been renewed; the lower you descend, the more you find the branches increase in size and diminish in number, till at length you reach the original and single stem.

CAROLINE.

But what is the difference between a subterraneous stem and a root?

MRS. B.

The structure of the root and of the stem is

in some respects different, and their functions totally so: the former merely sucks up nourishment from the soil and transmits it to the leaves; the latter is supplied with organs to distribute it, variously modified, to the several parts of the plant, the leaves, the flowers, &c.

There is a point or spot separating the stem from the root, called the neck, which may be considered as the seat of vitality. If you cut off the root of a young plant, it will shoot out afresh; if you cut away the stem, it will be renovated; but if you injure this vital spot, the plant will infallibly perish.

EMILY.

I think it should be called the *heart* rather than the neck of the plant, since it is so essential to its existence.

CAROLINE.

Is not the neck equally so? Animals will not survive decapitation any more than plants. But it is true the situation of the neck does not quite correspond with that of the animal frame, unless you denominate the roots the body of the plant, and the whole that is above ground the head.

MRS. B.

I do not think the huge trunk of a venerable oak would yield that title to its roots, and the extremity of its branches crowned with verdure would lay exclusive claim to the dignity of head.

The stems of plants are divided into two classes: those which grow internally, and those which grow externally. M. De Candolle distinguishes them by the characteristic appellations of *endogenous* and *exogenous*, a distinction first introduced by a celebrated French botanist, M. Desfontaines. We have no corresponding terms in English: in our country these two classes of plants retain the denomination given them by Linnæus, of *monocotyledons* and *dicotyledons*.

CAROLINE.

These are hard-sounding names, Mrs. B.: I hope their explanation will render them intelligible.

MRS. B.

I believe you will find no difficulty in understanding them. The class of plants whose stems grow internally, and are by us denominated monocotyledons, are distinguished by their seed, which, during germination, is converted into a thick leaf, yielding nourishment to the young plant until it is strong enough to suck it up from the soil. This leaf is called a cotyledon, and the epithet *mono*, which signifies *one*, implies that this class of plants have a single cotyledon.

The other class, whose stems grow externally, and are called dicotyledons, comprehends all those plants whose seeds in germinating split into two parts, forming two nutritive lobes or seminal leaves; and hence they bear the name of dicotyledons, which signifies two cotyledons.

EMILY.

I have seen lupins, peas, and beans germinate in this way; but do not recollect having observed any seed germinate with only one cotyledon.

MRS. B.

They are much less common in these climates, at least in plants of sufficiently large dimensions for their cotyledons to be observable.

There is a third class, denominated *acotyledons*, which have no cotyledons and no vascular system, such as fungi, lichens, &c.; but of these we shall not treat at present.

Let us first examine the structure of the stems of the monocotyledons or endogenous plants. Of this description are the date, the palm, and the cocoa-nut tree, the sugar-cane, and most of the trees of tropical climates.

Their stems are cylindrical, being of the same thickness from the top to the bottom; whilst those of Europe, you know, always become more slender and taper towards the summit of the tree, approximating to the conical form.

CAROLINE.

I thought that endogenous plants were those which grew in our own country, in opposition to exotics, or plants of foreign countries; but, by your account, it is just the reverse, for endogenous plants grow in countries most distant from us.

MRS. B.

You confound the word *endogenous* with *indigenous:* the latter signifies to grow within the country; the former is a French word, not yet introduced into the English language, signifying to grow internally, or within itself.

CAROLINE.

Within itself! How can the stem increase in size internally? One would think that the new layers of wood growing in the interior part of the stem would burst the external coats.

MRS. B.

The more the external coats are pressed by the new growing wood, the closer and more compact they become, and the greater the resistance they offer to the internal layers; till at length a period arrives when the outer coats are so hardened and distended as to yield no longer: the stem has then attained its full growth in horizontal dimensions, and offers a broad flat circular surface to view, which has scarcely risen in height above the level of the ground.

EMILY.

How singular a mode of growing! In this first stage it must resemble the stump of the trunk of a tree which has been cut down; but how does it grow up afterwards?

MRS. B.

The following spring, there being no room for a new layer of wood to extend itself horizontally, it shoots up from the centre of the stem vertically; fresh layers every year successively perforate this central shoot, till, from the innermost, it becomes the outermost layer of wood; hard, compact, and of the same horizontal dimensions as the base: the second period of growth is then completed; and thus the stem continues growing, for a certain number of years, horizontally, and then takes a sudden start upwards.

EMILY.

The stem then does not begin to rise until it is as large in circumference as at full growth. How I should like to see one of these broad flat stems!

MRS. B.

You may see them growing in hothouses; and though we have none in the open air in these climates, we have many smaller plants of the same description. Corn and all gramineous plants, the liliaceous tribe of flowers and bulbous roots, are all endogenous.

CAROLINE.

But lilies, tulips, and all flowers which spring from bulbous roots have long stems, thick at the lower end, and tapering towards the flower.

MRS. B.

You again confound the stalk of the flower with the general stem of the plant. Both flowers and leaves, with but few exceptions, have each a separate stem or foot-stalk: that of the flower is called by botanists a peduncle, or pedicel; that of leaves a petiole. These are perfectly distinct from, and independent of, the general stem of the plant. The stems of bulbous plants are contained within the bulbs as I have already informed you.

CAROLINE.

This mode of growing puts me in mind of the pushing out of an opera glass, the sliding cylinders of which are contained one within the other.

MRS. B.

The leaves and fruit of this class of plants grow from the centre of the last shoot, and form a sort of cabbage at the top of the tree, which, if you cut off, the tree perishes.

EMILY.

But what becomes of the bark of these trees? How does that resist the pressure of so many successive layers of wood?

MRS. B.

Endogenous plants have no real bark, the external coats of wood are so much hardened as to render such a preservation unnecessary.

EMILY.

But the palm and cocoa-nut tree, which I have

seen at Mr. Loddiges's hothouse, have a very rough external coat, greatly resembling bark.

MRS. B.

This is formed of the basis of decayed leaves. A circle of leaves annually sprouts from the rim of the new layer of wood; and, when they fall in autumn, leave these traces of their past existence. When a European woodcutter begins to fell a tree of this description, he is quite astonished at its hardness. "If I have so much difficulty with the outside," says he, "how shall I ever get through the heart of the wood?" But as he proceeds, he discovers that the trees of the tropical climes have tender hearts, if you will allow me the expression; this circumstance renders it very easy to perforate them, and makes them peculiarly appropriate for masts of vessels, pipes for the conveyance of water, and such like purposes.

These plants have usually no branches; but there is one species of palm-tree which shoots out two or three branches together.

The family of the gramineous plants, that is to say, the grasses and corn have a knot at the base of each leaf, whence the shoot grows.

CAROLINE.

I have observed that the straw of corn is hollow, but closed at certain intervals, forming externally a sort of ring; and it is from these rings that the leaves and branches shoot.

MRS. B.

The sugar-cane, which grows in this manner, is the largest of the gramineous plants.

Lilies are also of this description.

The Yucca of the tropics differs from our liliaceous plants only by having a longer stem; in these temperate climes vegetation has not sufficient vigour to develope all the energies of the plant, and the stem grows only laterally, never shoots upwards, but lies concealed in the bulb. Were it transplanted to a tropical climate, as soon as it had attained its lateral growth, it would shoot upwards in the manner I have described.

The structure of *exogenous plants* or *dicotyledons,* to which the trees of our temperate climes belong, is much more complicated. Here, then, are two reasons for our submitting them to a more accurate investigation.

The stem is composed of two separate parts: the one ligneous, the other cortical; in other words, it is formed of wood and bark.

The wood consists in the first place of the pith, a soft medullary substance, which occupies the centre of the stem, and is almost always of a cylindrical form. This soft pulpy body does not grow or increase in size with the tree, but retains the same dimensions it originally had in the young stem.

CAROLINE.

I thought that it rather diminished; for if you cut a young branch or stem, the growth of one season, the pith is very considerable, while little or none is to be discovered in the trunk of a full grown tree.

MRS. B.

The pith which fills the shoot of one season is scarcely perceptible in a large tree; the quantity, however, remains the same. Its dimensions may be contracted by the pressure of the surrounding coats of wood, which sometimes so condenses and hardens it as to prevent its being distinguished from them.

Some trees have a much greater quantity of pith than others; the elder-tree, for instance, abounds with it. The quantity of pith in the branches depends also upon their nature: if the branch is barren, it contains much more than if it is destined to bear fruit, but in the same individual stem or branch the quantity never alters.

The pith consists of cellular tissue. If this membrane be of a very fine texture, it is susceptible of extension as the branch lengthens; but if it be coarse, and the cells large, when the branch grows, it cracks and separates into parts. This is distinctly visible in a branch of jessamine, if you slit it open so as to exhibit the pith.

CAROLINE.

Here is one which we may examine. I will slit

it longitudinally: look, Emily, the pith is separated into parts, as if it had been forcibly torn asunder.

MRS. B.

It is the growth of the stem which thus rends the pith in pieces.

EMILY.

Then is it not destroyed and rendered useless?

MRS. B.

Yes; but not until it has fulfilled the purpose of its destination, which is to nourish the young wood during the first period of its existence.

EMILY.

It acts the part of a cotyledon or nurse to the young wood. But when it is become dry, what is to perform this office to the new wood which is annually formed?

MRS. B.

Every new layer of wood is lined with a layer of cellular tissue, which may be considered as the pith of the wood to which it is attached. These internal coatings not only separate the several layers of wood, but are also interwoven and incorporated with them, and may be seen in the form of rays, which appear to issue from the central pith, and proceed to the external layer of wood: these are called medullary rays; they are visible in wood, but are remarkably distinct in the root of the carrot.

CAROLINE.

But these fibres or rays, which appear all to proceed from the centre, cannot be continuous, since they originate annually in each fresh growth of wood.

MRS. B.

Very true; but they are so minute and so numerous, that the termination of those of one year's growth, and the commencement of those of the following year, cannot be distinguished. This gives them the appearance of being continuous; but were it really so, their distance from each other would increase in proportion as they diverged from the centre; yet you see in the carrot they are as close, and consequently much more numerous, in the external layers of wood, than in those nearer the central part. In one sense, indeed, they may be considered as continuous; as it is conjectured that the growth of the new wood originates from the extremities of the medullary fibres of the preceding year: this would tend to give regularity to the distribution and direction of the successive rays, and an appearance of continuity. A succession of these horizontal rays, perfectly regular, form vertical planes along the stem, which may be tolerably well represented by those circular brushes which are made to clean the inside of bottles.

EMILY.

The wood of exogenous plants, growing externally, has not the same difficulties to encounter as that of endogenous plants.

MRS. B.

The difficulty is rather reversed than diminished, the pressure being from the external upon the internal parts. The first layer surrounding the central pith grows freely during a twelvemonth, but the following year it is enclosed by a new layer; and notwithstanding the accession of nourishment it receives from the roots, and the additional space it would, if unconfined, occupy, it is pressed and squeezed by the new layer into a narrower compass than it occupied the preceding year. In this distressing situation, what is to be done? Compelled to yield laterally, it makes its way where there is no pressure; that is to say, vertically: thus the stem grows in height at the same time that it increases in thickness. The first layer of wood having, therefore, found a vent for that new portion of its substance which could not be contained in the contracted space in which it was confined by the growth of the second layer, this portion grows freely during the second year; when a third layer shooting up around and compressing the second, this in its turn escapes from bondage, but, rising vertically, it encloses and confines the first layer.

CAROLINE.

The second layer from the prisoner becomes

the gaoler; but its prisoner does not excite my commiseration, for the first layer, having learnt how to escape, doubtless profits by its experience, and rises above the fetters with which it is encircled.

MRS. B.

Yes; the first layer thus makes a shoot upwards every year, and the new layers follow its course in regular succession. This mode of growing, you must observe, renders the form of the stem conical, the number of layers diminishing as the stem rises.

These layers of wood attain a state of maturity, when they become so hard by continued pressure as to be no longer susceptible of yielding to it. Previous to this period, the layers bear the name of *alburnum*, signifying white wood, for wood is always white until it reaches this degree of consistence. The length of time requisite to attain a state of maturity varies extremely, according to the nature of the wood. In some trees five years are sufficient for this purpose; in others ten or twenty are necessary; and the Phyllyrea requires no less a term than fifty years to convert its alburnum into perfect wood. When once the first layer has attained this point of maturity, the others naturally follow in succession, according to their respective ages.

EMILY.

But are those dark-coloured woods, such as mahogany and rose-wood, ever white?

MRS. B.

Yes; and, what is still more remarkable, ebony, a wood which is completely black, is white until it has attained this state of maturity. Here is a small piece of a branch of ebony cut transversely: you see that the interior parts are perfectly black, and are surrounded by a ring of white wood or alburnum. The difference between the alburnum and perfect wood is less marked in woods of a lighter colour, but it is always sufficiently so to be distinguishable. Look at this trunk of chesnut-tree, which has been recently cut down with a saw.

CAROLINE.

I not only see plainly where the perfect wood is separated from the alburnum, but I can distinguish every layer of wood. I follow them in imagination in their successive shoots upwards to extricate themselves from the pressure of the new layers, by counting the number of layers at the base of the tree; then, Mrs. B., shall I be able to ascertain its age?

MRS. B.

Yes; and you may do more: for if you take the trouble to count the number of layers at each end of one of those pieces of wood which have been sawed into logs for fuel, you will learn how many years that portion of the tree was in growing.

EMILY.

There are thirty layers at one end, and twenty at

the other; consequently the tree must have been ten years growing the length of this log. I little thought I could ever have taken so much interest in a log of firewood.

MRS. B.

However mean or common-place may be the purposes to which we apply the works of Nature, when studied in a philosophical point of view, they are no less objects of interest and admiration.

The annual layers of wood are distinguishable not only by their different degrees of hardness and density, but also by their being separated by layers of the cellular system; so that, when you examine the trunk of a tree, you perceive zones of woody fibre and zones of the cellular system.

EMILY.

Can the age of endogenous plants be ascertained in the same manner?

MRS. B.

No; the annual layers of wood are not sufficiently distinct from each other.

CAROLINE.

But the rings annually formed by the vestiges of leaves is a still better record of their age, for it is not necessary to cut down the tree in order to ascertain it.

MRS. B.

For a certain period they may answer this purpose; but these vestiges are obliterated by time, and in an aged tree are no longer distinguishable towards the base of the stem.

Bark.—The vegetation of the bark is precisely the inverse of that of the wood; that is to say, it is *endogenous*, its layers growing internally like those of the palm-tree: the new soft coat of bark therefore lies immediately in contact with the new soft layer of wood.

EMILY.

But if a fresh layer of bark grows every year, why is the bark so much thinner than the wood? I should have supposed that they would have been of equal dimensions?

MRS. B.

The outer coats of bark, when they become too hard to be further distended by the pressure of the internal layers, crack, and, becoming thus exposed to the injury of the weather, fall off in pieces: it is this which produces the ruggedness of the bark of some trees. In others, the rind, though smooth, peels off, after cracking, like that of the cherry, the birch, and particularly the plane-tree. Those trees whose external coat of bark is least liable to peel off, such as the oak and the elm, become more scarred and rugged, in proportion as the tree grows older, and is longer exposed to the action of air, water, insects, and parasitical plants: sooner or later these various causes effect the destruction of the outer bark; and the other layers, as they become external and exposed to the same sources of injury, experience, in due course of time, the same fate; whilst the layers of wood, being protected and sheltered by the bark, vegetate in security.

CAROLINE.

Yet it is not uncommon to see the trunks of very old trees in a state of total decay, whilst the bark remains uninjured.

MRS. B.

That is the case when the wood is, by any accidental circumstances, exposed to the inclemencies of the weather, which it is not calculated to resist. This happens, sometimes, by the lopping or breaking off of large branches, considerable pieces of bark falling off, or any circumstance by which the rain can gain admittance to the wood.

There are some trees whose bark is of so elastic and yielding a nature, that it does not harden for a considerable number of years. The bark of the cork-tree, for instance (which is the part commonly called *cork*), does not begin to harden till after the age of seven years: care is taken to strip it off for the use of the arts before that period. The bark of the plane tree is, on the contrary, of so hard and inflexible a texture, that it cannot expand, but splits and falls off every season. These two species of trees, the cork and the plane tree, form the two extremities in the scale of varieties of texture in the nature of bark. The cuticle or external coating of bark is not confined to the stems and branches, but spreads itself over the leaves, and every part of the surface of the plant which is of a green colour.

EMILY.

But the bark of trees is not of a green colour, Mrs. B.?

MRS. B.

Recollect that the cuticle is an envelope, which lasts seldom more than a twelvemonth. In those parts of a plant which are of longer duration, such as the stem and branches of trees, the cuticle decays and peels off; and its place is supplied by the epidermis, a coating formed by the desiccation of the external part of the cellular tissue which has been left exposed to the air. The epidermis, therefore, is not green.

Aquatic plants form the only exception, these having, properly speaking, no epidermis.

If you pass a silver wire or *blade* completely though the bark of a tree, the new internal layers, as they are annually formed, will gradually push it outward, till at length the internal coat becoming external, the wire will fall off.

CAROLINE.

That is, no doubt, the cause why inscriptions on the bark of trees are, in the course of time, effaced: the new bark does not grow over them it is true, but growing under them, the inscription becomes distended, and when the bark gives way, it will most readily split and fall off where the inscription has already wounded it.

EMILY.

If, however, the inscription be made so deep as to penetrate the layers of wood, the new layers of bark, instead of injuring, will preserve it.

CAROLINE.

But of what use will be its preservation, whilst it is so buried as to be totally lost to the sight?

MRS. B.

Buried treasures are sometimes brought to light. Adamson relates, that, in visiting Cape Verd in the year 1748, he was struck by the venerable appearance of a tree, 50 feet in circumference. He recollected having read in some old voyages an account of an inscription made in a tree thus situated. No traces of such an inscription remained, but the position of the tree having been accurately described, Adamson was induced to search for it by cutting into the tree, when, to his great satisfaction, he discovered the inscription entire, under no less a covering than three hundred layers of wood.

CAROLINE.

Three hundred years, then, had elapsed since the inscription had been made! How much he must have been gratified by the discovery! — But did not his venerable tree suffer from such deep wounds?

EMILY.

Probably not: for, according to the size of the tree, though he cut so deep, he was still far distant from the centre.

MRS. B.

The centre is not the most dangerous part: on the contrary, the vital part of the stem is situated between the young layers of wood and those of the bark; or perhaps the vitality may be exclusively confined to the inner coat of the bark: for if the young layer of wood be destroyed by frost, the tree suffers but little; whilst, if the inner coat of bark be frozen, the plant infallibly perishes. In the trunk of a tree which has been cut down, it is very easy to trace the effect of frost on any layer that has been injured by it, the wood appearing withered and wrinkled. Mr. De Candolle observed a frostbitten part of this description in a tree cut down in the forest of Fontainbleau in the year 1800; and, by counting the superincumbent layers of wood, he ascertained that it must have happened in the year 1709, one which was remarkable for the severity of the frost.

EMILY.

But since the layers of wood grow with so much regularity, whence come those knots and waving lines, which constitute the beauty of polished wood?

MRS. B.

If the sap, in rising through the young wood, meets with any casual obstruction to its passage, it naturally accumulates in that spot, and forms what is called a knot. This consists of distended vessels, containing a magazine of food, which gives birth to a germ or shoot; but it frequently happens that, before this germ has attained strength to force its way through the bark into the open air, a new layer of wood rises over and encloses it. Sometimes it is only temporarily buried; and the following season it acquires sufficient vigour to break through its prison. Thus, if the shoot go on annually forcing its way through the wooden wall which rises up to oppose its progress till it reaches the surface of the stem, it becomes the origin of an external shoot or branch. If, on the contrary, it is exhausted by this series of struggles, it perishes; and leaves, in memorial of its efforts, the knots, waves, and streaks, which embellish its tomb. This shoot, which had increased in size whilst traversing the several layers of wood, as soon as it grows externally, diminishes as it protrudes in the air, being thickest at the stem, and tapering towards its extremity; so that a shoot, if traced from its origin, exhibits the form of a double cone, the base of which is at the surface of the stem.

EMILY.

But whence did this shoot derive its origin?

The accumulation of sap can merely favour its growth, but cannot have given it existence.

MRS. B.

This is a question not very easily answered; but the opinion most prevalent among botanists is, that germs or latent shoots exist throughout the stems and branches of plants, and that those only are brought into a state of active vegetation which are fully supplied with food.

CAROLINE.

Do the stems and branches of exogenous plants grow like their roots, merely at their extremities?

MRS. B.

No; they increase throughout their whole length. If you make marks at certain distances on a root, you will find that these distances are not altered by growth; but if you make similar marks on a stem or a branch, the distances will increase, showing that it grows in its whole extent.

EMILY.

It must be so, since a new layer of wood grows annually at the base.

And pray, through what part of the stem does the sap rise?

MRS. B.

That is a question which has been long and much disputed. Some naturalists have maintained the opinion that it ascended through the pith: others, that it rose through the bark: and they have reciprocally proved each other to be mistaken in their conjectures. A third road was, therefore, sought for; and, by colouring the water with which a plant was watered, it has been traced within the stem, and found to ascend almost wholly in the alburnum or young wood, and particularly in the latest layers.

CAROLINE.

That is very natural. The perfect wood has in a manner finished its active career: it can itself acquire but little nourishment; and its indurated texture would be ill adapted to the conveyance of the sap, whilst the young layers being in the full vigour of growth, and their cellular system flexible and elastic, are much better calculated to transmit it; besides, it is in these, you say, that the young shoots take their origin.

MRS. B.

The sap does not impart nourishment to the plant during its ascent: it is therefore more probable that its rising through the new wood is owing to that being softest and most permeable. By means of the coloured medium I have mentioned, it was observed that the sap naturally ascended in straight lines, but that, if it encountered any obstacle, it could move obliquely, or even spread itself laterally.

A great variety of experiments have been made in order to ascertain the degree of velocity with which the sap rises; but as the rapidity of its ascension depends in a great measure upon the means which the plant has of parting with it by exhalation, we cannot well follow its progress without having previously made acquaintance with the excretory organs of plants — the leaves, whose office it is to exhale that portion of the sap which is superfluous.

CAROLINE.

The whole of the sap then is not required for the nourishment of the plant?

MRS. B.

That nourishment is a more complicated operation than you are aware of: all the water which enters into the plant is not retained by it; part of it passes through the leaves into the atmosphere, and the atmosphere, in its turn, contributes to the nourishment of the plant. But we must not anticipate; and, at our next interview, we will examine the structure and agency of the leaves of plants.

CONVERSATION IV.

ON LEAVES.

MRS. B.

There is nothing more beautiful in the vegetable creation than the gradual formation and developement of a leaf. It consists of the flattened expansion of the fibres of the stem from which it shoots, connected together by a layer of cellular tissue called the pabulum, and the whole is covered by a delicate coating of cuticle, which is almost always of a green colour. A plant may, indeed, be considered as a continued series of these fibres, sometimes closely bound up in the form of stems, at others spread out into that of leaves.

CAROLINE.

Yet surely, Mrs. B., there are many parts of a plant which can neither be referred to leaves nor stems? The blossom, the fruit, and such occasional appendages as thorns and tendrils, cannot come under either of these denominations?

MRS. B.

I beg your pardon: they all originate in leaves. Even the seed, when first ushered into life, comes cradled in a folded leaf; but as in assuming the form of a seed-vessel it loses that of a leaf, we must not allow it to encroach upon the present subject of our conversation, that of leaves properly speaking, which retain their original form throughout the whole of their transitory existence.

CAROLINE.

Well, it must be confessed that this borders on the marvellous; but I shall take it on your authority till the time comes for you to explain it more fully.

MRS. B.

It rests upon much better authority than mine: it is sanctioned not only by the opinion of Mr. De Candolle, but also by the celebrated Mr. Brown, who was the first to develope this theory in England. In Germany, so long as thirty years ago, the venerable poet Goëthe wrote a small treatise on the metamorphoses of plants; and if this little work has not met with the attention it deserves, it is probable, that, being written by a poet, it has been considered rather as the effusion of an ardent imagination, than as the deductions of a philosopher. But, whatever be the changes which leaves may undergo, it is our present business to treat of them in their state of leaves.

If, when a leaf shoots, the fibres which attach it to the stem or branch spread out immediately, the leaf is termed *sessile* or continuous; for it cannot be separated from the stem without the fibres being torn asunder; the leaves of corn, grasses, and all gramineous plants, are of this description.

EMILY.

But it is much more common for leaves to be attached to the branch by a foot-stalk.

MRS. B.

With exogenous plants it is; and the trees and shrubs of our temperate climate are almost all of that class. Such leaves are said to be articulated: the fibres when they first separate from the stem remain bound together, forming the *petiole* or foot-stalk; thence they expand in numerous ramifications, constituting the ribs of the leaf. Let us now examine this leaf of a horse-chesnut: I cut it transversely at its base, and you may perceive with the naked eye the larger vessels which convey the sap into the leaf. At the other extremity of the foot-stalk they are also visible. They are five in number, corresponding with the five leaflets of which the horse-chesnut leaf is composed.

The fibres of leaves spread out in various directions: the principal one, dividing the leaf from the base to the summit, is called the dorsal, or mid-rib; others branch out from this laterally; and a third class consists of still smaller ramifications issuing from these last: they all terminate at the surface of the leaf by a pore called *stoma*, a Greek word, signifying mouth.

CAROLINE.

These are, no doubt, the exhaling pores which send off the superflous moisture?

MRS. B.

Yes; but we must patiently labour through a forest of foliage, before we can return to the physiological operations of the plant.

Leaves are usually divided by botanists into five classes, according to the direction of their ribs: —

First, the pennated are those in whichthe smaller ribs expand from the principal rib like the feathers of a quill: the leaves of the pear and the lime-tree are of this description.

The second class is palmated. In these, the ribs diverge from the petiole like the fingers from the palm of the hand, as you see in this vine-leaf. They are not, however, always five in number, varying not only in different plants, but sometimes in different leaves of the same individual.

The third class is called target-shaped, or *peltate*, being shaped like a buckler; such is the nasturtium.

The fourth class is *pedatum*, having the form of the foot: the hellebore is of this class.

The fifth class has simple ribs, proceeding from the base to the extremity of the leaf; corn, grasses, and all the gramineous tribe are comprised within it. These leaves are always sessile.

The contour, or external form of the leaf, is of much less importance than the direction of its ribs. The indentures, or teeth of leaves, are formed by the termination of its ribs.

In the gramineous tribe, the leaves are smooth at the margin, and have no indentures; the ribs run on each side along the margin like a small seam, and terminate at its pointed extremity, whence all the exhalations take place.

When the indentures of some leaves reach so far as half-way down, they are said to be pinnatifid; and when the leaves, though separate, grow from one foot-stalk, so that one of them cannot fall off, or be separated from the other without being torn asunder, the leaf is said to be dissected.

CAROLINE.

There are a great variety of leaves of this description: the rose, the acacia—

MRS. B.

No; these are compound leaves, and differ from the dissected by being articulated, each leaflet having a separate foot-stalk, which, when the leaf dies, detaches the leaflet from the general footstalk, and they fall separately.

At the base of the foot-stalk of compound leaves there generally grows a small organ, called *stipula:* it consists of two accessary leaves, as you see here in the rose-leaf, the willow, and indeed in most exogenous plants. Sometimes the stipula is attached to the foot-stalk, at others to the stem: it withers easily, and often falls off before the other leaves; for which reason it is not always to be met with on branches of a certain age. In this branch of rose-tree you see that there are stipulæ to all the younger shoots, while the older ones have already lost them. In the pea the stipula is larger than the common leaves. [See Plate I.]

When the ribs of leaves are expanded upon the same plane, the leaf is thin; in succulent plants, which retain moisture and evaporate but little, the cellular tissue, which connects the vessels of the upper and lower surface of the leaf, is thick and fleshy.

The two surfaces of a leaf generally differ in appearance: in the upper surface the ribs are the least prominent, and the leaf is consequently the smoothest, and of the deepest green. The under surface is more hairy, and abounds with *stomas* or pores; the upper has fewer, or is sometimes wholly deprived of them, excepting in aquatic plants, whose leaves float on the water; their upper surface being alone exposed to the air, are alone supplied with *stomas*.

But whether the two surfaces be similar or not, it is very certain that their functions are different; for if you reverse the leaf of a plant, and prevent it from resuming its natural position, it will wither and die.

EMILY.

But corn and grasses grow vertically, Mrs. B., and can scarcely be said to have an upper and an under surface; though, it is true, they are greener and smoother on one side than on the other.

MRS. B.

All the gramineous family have a more equal distribution of pores on either surface; for growing nearly erect, and being therefore equally exposed to the air, each surface can probably perform the same functions, and these plants can bear the reversion of their leaves better than any other.

Floating aquatic plants, on the contrary, having no pores on their lower surface, infallibly die if they are reversed without power of resuming their natural position.

CAROLINE.

It would be superfluous for aquatic plants to be furnished with pores on their under surface, since they could not evaporate into water.

MRS. B.

Nor can they elaborate the sap without exposing it, by means of the pores, to the atmosphere: but we must complete the anatomical examination of the structure of the leaf, before we enter upon its physiological functions.

The first appearance of leaves which the young plant puts forth on the germination of the seed is formed by the lobes of the seed itself, which we have already noticed under the name of cotyledons.

EMILY.

I have often observed them in lupins, when they first shoot above ground, and wondered that the tiny plant should be able to supply food to such thick substantial leaves.

MRS. B.

It is, on the contrary, these leaves which yield their substance to the tiny plant; and as soon as they have completed this function, and the whole of their pulpy nutriment is consumed, they wither and fall off.

But all cotyledons are not of a succulent nature: some are thin, like other leaves, and are more commonly called seminal or seed leaves.

EMILY.

How, then, can they feed the young plant?

MRS. B.

By immediately elaborating the sap, which the nascent root draws up from the soil. Seminal leaves are furnished with stomas for this purpose, while fleshy cotyledons have none; in the latter, the conversion of the cotyledons into leaves is but very imperfect: they frequently remain under ground, and do not assume either the form or colour of a leaf.

EMILY.

The cotyledons of peas and beans are of this description; in those of lupines the conversion is more complete, though they remain succulent.

CAROLINE.

Since the fleshy cotyledons have no stomas, I know not what they have to do in the open air: merely acting the part of a magazine of food, they are more at hand to supply the young plant with it under ground than above it.

EMILY.

But is it not wonderful that a young plant should be able to absorb sap, and elaborate it from the first moment of its existence?

MRS. B.

Not more so than that a young chicken should pick up grains of corn as soon as it has thrown off its egg-shell. Nature has probably given more firmness and stability to the roots of plants, which are obliged immediately to provide their own food, in the same way as she has to the beaks of young birds. The embryo plant has often another resource, but which does not belong to our present subject.

The first regular leaves which expand are called primordial, and are not unfrequently of a different character from the common leaves of the plant.

When the leaves shoot very near the ground, so as to appear to grow from the roots, they are called radical leaves; they sprout, however, from the base of the stem, for roots never give birth to leaves.

Bracteæ or floral leaves are, on the contrary, leaves peculiar to some plants, which grow very near the flower, and are often mistaken for blossom, not being always of a green colour. The Hydrangea, for instance, has a great abundance of pink or lilac bracteæ, which I doubt not but that you have supposed to be the flower of that plant.

CAROLINE.

Are then those beautiful blossoms of the hydrangea not its flowers?

MRS. B.

To a superficial observer they bear, it is true, a much greater resemblance to flowers than to leaves; but, if examined attentively, you will find they possess few of the regular organs of the flower, and could produce neither fruit nor seed.

EMILY.

Is there any other example of coloured leaves which are bracteæ instead of blossoms?

MRS. B.

A great number. The lime-tree shoots out a profusion of bracteæ of a pale yellow colour; and I doubt not but that you have confounded them with the blossom which lies concealed beneath them. The coloured leaves of the *red-topped Clary*, which exhibit various tints of red, purple, and green, are also of this description. There are many bracteæ which do not differ either in colour or form from the other leaves of the plant, and are distinguished only by their situation with regard to the flower. Such is the tuft of leaves on the summit of the flower called the crown imperial, and that which grows from the top of the pine-apple: the scaly covering of this fruit consists also of the remnants of degenerated bracteæ.

CAROLINE.

And pray, Mrs. B., is not the scaly cone of the fir-tree of the same nature? I have often observed it when the seeds have fallen out, and it wears the appearance of an aggregation of short, thick, stiff leaves, forming a cone of cells somewhat resembling a bee-hive.

MRS. B.

You are quite right; except in calling the fruit which is lodged in these cells seeds: their botanical name is achenium.

Both the fir-cone and the pine-apple are aggregated fruits, separated by bracteæ; but in the succulent pine-apple, almost all vestiges of the intervening bracteæ are obliterated.

EMILY.

When the crown of the pine-apple is pulled

off, the summit of the fruit, I think, exhibits some marks of cells formed by bracteæ.

MRS. B.

That is true; and they are, you may have observed, empty: the pressure of the base of the crown having prevented the fruit from growing, the bracteæ are not wholly obliterated.

Leaves are arranged on the stem in a great variety of ways: sometimes in opposite pairs, and the successive pairs crossing each other at right angles; at others, several leaves shoot from the same spot, and spread out in a circle. They sometimes alternate on the stem, and appear irregularly scattered; but Nature allows nothing to be scattered by chance: upon a careful investigation, order and method will be discerned in the minutest of her works; and, in the arrangement of leaves on the stem, she has been studious to prevent their covering each other too closely, both light and air being required to enable them to perform their functions.

EMILY.

Is it not surprising that Nature should have bestowed so much pains upon so insignificant a part of the creation as a leaf; which, however beautiful and curiously constructed, lasts but a season, and is then scattered by the first blast of wind, and trodden under foot?

MRS. B.

Not until it has performed the part which Nature has assigned to it; and when you are acquainted with the importance of its functions in the vegetable economy, you will probably be induced to treat it with more respect.

CAROLINE.

Leaves, when they first shoot, are generally enclosed in small scaly buds, evidently designed to protect them from inclemency of weather. Now these scales differ totally in form and appearance from the leaves they shelter; and I think, Mrs. B., that you would be at a loss to derive them from the same origin?

MRS. B.

Nothing more simple. All leaves begin to shoot without any external covering; but when, in early spring, they quit the protecting branch in which they were embosomed, to enter into the cold region of the atmosphere, they are chilled and checked in their growth, and, instead of expanding in the natural form, they contract, harden, curve inwards, and are finally transformed into a species of scales, which serve to protect the internal leaves: under so friendly a covering, these vegetate freely. In the mean time the season advances, the atmosphere acquires heat, and the young leaves, having been protected from its former inclemency, are cherished and developed by its genial influence.

EMILY.

What a beautiful provision for the security of a leaf!

MRS. B.

If you follow up the developement of the bud of the ash or the maple, you will observe that the external scales are short, hard, reddish, and rather hairy. In proportion as they are more internal, they become membranaceous, pale-coloured, and elongated; small rudiments of leaves then appear at their extremities; and these, shortly after assume the form of leaflets, so very different in shape and structure from the external scales, that it is difficult to conceive they have had the same origin.

EMILY.

The more feeble and delicate the leaves of a plant are, the greater, I suppose, will be the number of those which degenerate into scales; therefore, the thicker and warmer will be the covering for the leaves which are ultimately to be developed.

CAROLINE.

And these, being of the same delicate texture, require such an additional clothing. What an admirable effect produced from so simple a cause!

MRS. B.

These scaly leaf-buds are not universal, some leaves being of so hardy a nature as not to require a covering, especially when growing in a warm climate: they are then said to grow naked; but being closely folded or rolled up in a small compass when first they shoot, they wear the appearance of a smooth bud without scales.

The horse-chesnut, in its native climate of India, unfolds its young leaves to the general atmosphere, without risk of their suffering from exposure; while, in this colder country, many successive leaflets are arrested in their growth, and condemned to degenerate into scales. If you examine the buds on this branch, you will see what numbers have changed their form, and are reduced to play a subordinate part in the system of vegetation.

The scales of some buds are formed from the rudiments of stipulæ; others derive their origin from petioles or footstalks; which, instead of growing long and slender, expand and assume the form of scales, and envelope the embryo shoot. The buds of the walnut and the pear are formed from stipulæ.

EMILY.

I have often examined these buds with great interest, and admired the ingenious manner in which the leaves were so closely packed, in order to be contained within them. Do the same buds produce both leaves and flowers?

MRS. B.

Buds vary in this respect not only in different plants, but sometimes even in the same individual:

some sprouting into flowers and fruits; others into leaves only, and branches; and there are buds of a third description, which develope both fruit and leaves. The first kind is full and round; the second, smaller and more pointed; and the third, both in size and shape, forms a medium between the other two.

CAROLINE.

How essential it must be for a gardener to be able to distinguish these buds! For if, in pruning a tree, he were to lop the branches which contained most of the fruit-buds, and retain those which had more leaf-buds, he would have a very poor crop of fruit. Are these three species of buds common to all trees?

MRS. B.

No; the buds of the horse-chesnut, which are so large, scaly, and glutinous, are all of the mixed kind; those of the apple and pear are of the two distinct species.

Endogenous plants, or monocotyledons, scarcely ever produce more than one single bud annually; the cabbage of the palm-tree is its bud, and the leaves and flowers are folded within it. The cocoa-nut and date trees develope their flowers and foliage in the same manner.

Bulbous plants (the endogenous plants of our temperate climate) are of the same description. I have already observed, that their stem is contained within the bulb; but you have yet to learn that this bulb is in fact the bud or cabbage, containing not only the stem, but also the leaves and flowers. The scales formed for the rudiments of undeveloped leaves are particularly distinct in bulbous roots, especially in the onion.

EMILY.

Thus then a lily, a tulip, or a hyacinth, are all contained within their bulbs, which we have been accustomed to consider merely as their root. But these flowers have each a stem, Mrs. B., independently of that which you say remains undeveloped within the bulb.

MRS. B.

The shoot which you consider as a stem, is the peduncle or footstalk of the flower, not the stem of the plant; the leaves which grow from the stem shoot from beneath the footstalk. The bud or cabbage of the palm-tree, when developed, shoots up a footstalk on which the flower expands, while the leaves spread out at its base. The difference between these plants of the tropical and of the temperate zones is, that the stem of the palm-tree being developed, the cabbage is situated at its summit; whilst, in our more temperate climate, the vegetation of the bulbous plants is not sufficiently vigorous fully to develope their organs, and the stem remains in a latent state within the bulb.

CAROLINE.

How can those plants bud whose stem and branches die annually, such as dahlias, pæonies, and China asters, &c.?

MRS. B.

The new stem and branches shoot from a bud formed at that vital spot called the neck of the plant; in perennials the stem dies down to this spot, but if that perishes, the whole plant dies. It is situated either on a level with, or rather below, the ground; and the bud being but little exposed to the weather, is not provided with the same warm covering as most of those which sprout in the air.

CAROLINE.

That is to say, being protected by their situation, the first rudiments of their leaves do not degenerate into scales.

And pray, do all these various kinds of buds originate in some little accumulation of sap in the stem, in the manner you have described to us?

MRS. B.

This accumulation of sap is the origin of a bud, so far only as it enables a germ or embryo shoot to grow, by affording it an ample supply of food.

They commonly shoot at the articulation of a leaf, because the branching off of the vessels offers some little impediment to the flowing of the sap; a small portion of it is arrested in its course, and forms a deposition of food, which a neighbouring germ quickly applies to its own use, and is thus ushered into life.

EMILY.

Such germs must exist then in a latent state in every part of the stem, and wait only for means of sustenance to make their appearance externally.

MRS. B.

In all probability; for wherever there is an accumulation of sap, a germ is sure to be developed.

EMILY.

When I plant a slip of geranium, I take care that it should have at least one leaf; because I know by experience, though I was quite ignorant of the cause, that the shoot would spring from the articulation of the leaf.

MRS. B.

In geraniums there is also another species of articulation, consisting of knots in the stem, which answers the same purpose, by interrupting in some measure the circulation of the juices, and affording a little supply of stagnant sap. Pinks and carnations, reeds and rushes, the stems of corn and grass, are all intersected in a similar manner.

CAROLINE.

Then when carnations and pinks are propagated by layers, the shoots take place at those intersections of the stem.

EMILY.

Excepting the geranium, the leaves of all these plants are, I believe, sessile; the intersections in the stems must therefore supply the place of articulations in furnishing the buds with food.

MRS. B.

Precisely so. They belong to the class of endogenous plants, whose leaves are more rarely articulated than those which are exogenous. These intersections, however, not unfrequently occur in the latter, as with the geranium, the vine, and several other dicotyledons.

You may recollect my observing, that a number of years frequently elapsed before the buds formed on the stem or branches of a tree attained sufficient strength to force their way through the successive layers of new bark which annually enclosed them; while others vigorously pushed through this barrier the first year. Buds usually begin to be formed in the month of August, and remain in a latent state during the winter, when they are commonly called *eyes:* the following spring they shoot; but they cannot properly be called buds till the scales are formed by the degeneration of the external leaves of the shoot.

It is heat which determines the period of budding of a plant. A branch turned towards the south, or introduced into a greenhouse, will shoot long before the rest of the tree; the budding begins to appear first near the extremity of the branches where the wood is most soft and tender.

EMILY.

I should have imagined that the base of the branch which the sap first reaches would have budded earliest.

MRS. B.

In the larch, and many other trees whose branches are equally hard throughout, this is the case; but the superior facility of piercing through the tender part of a branch more than compensates for the earlier supply of food.

The scales of buds are often coated with a sort of glutinous varnish, which resists moisture; some are lined with a species of down or fur, to preserve the internal shoot from cold.

CAROLINE.

But can down or fur result from the degeneration of leaves? Such a beneficent provision for the protection of the shoot would seem to indicate, that the bud of a distinct organ is specifically designed for that purpose.

MRS. B.

Such is no doubt equally its destination, whether it originate in the abortion of another organ, or whether expressly created for that purpose: nor is it difficult to refer the formation of down or varnish to the same origin. The scales of buds probably absorb from the sap only a portion of what was destined to nourish them had they been developed into leaves, and the remainder may be converted into a species of glutinous resin or varnish. The rudiments of leaves, when examined in the bud before it is developed, wear the appearance of small filaments of cotton, which, when spread out, exhibit the minute skeleton of a leaf.

It would be difficult to suggest a mode of folding or rolling which Nature has not adopted in enclosing these embryo leaves in the bud. They are sometimes, as those of the vine, folded like a fan; others are doubled from the top to the bottom; others folded down the middle; some are laid one within another; others closely packed side by side; and there are an equal number of modes of rolling them up in the buds.

In some plants the petioles or footstalks retain the nourishment they should transmit to the leaves, so as to prevent the latter from being fully developed; they remain therefore in an embryo state; the petiole, in the mean time, gorged with nutriment, becomes thick, corpulent, and clumsy, flattening as it expands, and wears rather the appearance of a leaf than of a stalk. The acacia of New Holland has this singular conformation.

CAROLINE.

I have seen tropical plants in hothouses of this description: the prickly fig is, I believe, one of them. But how do these leafless plants disburden themselves of the superfluoüs moisture which leaves exhale?

MRS. B.

The dilated petioles, which usurp the place of leaves, perform also, though but imperfectly, their functions, and have pores adapted for that purpose: they are not, however, leaves, any more than the tail of a kangaroo is a leg, or the trunk of the elephant an arm, though they respectively perform the office of these members. When common organs assume in certain species an uncommon form, they may be useful for purposes different from those for which Nature originally designed them; but they should not on that account obtain the name of the organ they but imperfectly imitate.

Leaves are usually deciduous, that is to say, last but one season: there are but few exceptions of plants whose leaves last two, three, and sometimes as long as four years. Evergreens change their leaves annually, and the plant remains green only because the young leaves appear before the old ones decay.

EMILY.

Is it not singular that the leaves of evergreens

should wither and fall in the spring, when the weather becomes warm, the sap most abundant, and vegetation in full vigour?

MRS. B.

A leaf withers when the vessels which should bring it nourishment are no longer capable of performing that function. In autumn, the vessels of the petiole become obstructed by a deposition of hard matter, which disables them from transmitting sap, and, being no longer moistened by the passage of this fluid, they dry up and wither; while the pabulum of the leaf, consisting of an expansion of the cellular system, which is of a soft, moist nature, preserves the leaf some little time after the petiole has ceased to perform its functions.

CAROLINE.

Like an animal deprived of sustenance, it feeds on its own fat, before it perishes.

MRS. B.

The circulation of sap in evergreens being more uniform throughout the year, the deposition of hard matter does not obstruct the passage of the sap till towards the spring, when the vigorous sap is directed towards the buds, and the old leaves drop off as the young ones expand.

The petioles of some leaves, such as the aspen and the poplar, are flattened, and adhere less firmly to the stem; hence they tremble at every breath of wind, and fall off more readily than those of a cylindrical form.

With regard to the most important functions of the leaves, the chemical changes they operate upon the sap, we must reserve them for our next interview, which I propose to dedicate to the examination of the sap, and the interesting part it performs in the vegetable system.

CONVERSATION V.

ON SAP.

MRS. B.

Now that you have made acquaintance with the *root*, the *stem*, and the *leaves*, we may proceed to trace the sap in its ascent through these several organs, observe the various transformations it undergoes in the leaves, and, following it in its descent, examine the manner in which it feeds and restores the several parts of the plant.

CAROLINE.

This seems to me to comprise the whole history of vegetation.

MRS. B.

In a general point of view it does, but we shall yet have many details to enter into; besides what I have hitherto said relates only to the nourishment of plants: their re-production is of no less importance, and we have not yet ever alluded to the flower, the most distinguished and beautiful of their organs, and that in which the seed originates.

EMILY.

But this sap, Mrs. B., which I imagined to be diffused through the plant as it rose, seems to be disposed of in a very different manner: part you say is exhaled by the leaves, and part descends through the bark; what then remains to nourish the plant?

MRS. B.

All that is necessary for that purpose is selected and retained. If you consider that the sap which rises in the roots consists simply of water, holding in solution a variety of crude ingredients, such as lime, silex, magnesia, soda, and potash, you will acknowledge that something more is required than the mere diffusion of this heterogeneous fluid through the plant in order to nourish it. The sap traverses the stem, rising, as I have already said, through the alburnum, and some small portion of it through the perfect wood. A great variety of experiments have been made, with a view of ascertaining the degree of rapidity with which the sap ascends. M. Bonnet raised some plants in a dark cellar, in order to blanch their stems, that he might be able to trace the ascent of the coloured water with which he nourished them. He found that this tinted sap rose only four inches in two hours; but the plants, owing to the disadvantageous circumstances under which they were cultivated, were weak and sickly; in subsequent experiments on more healthy plants, the sap was seen to ascend three inches in the course of an hour. Some time afterwards Mr. Hales immersed a fresh cut branch of a vigorous pear-tree in a tube full of water, and found that the sap rose in it eight inches in six minutes.

EMILY.

And how do you account for so remarkable a difference in the result of these experiments?

MRS. B.

Chiefly from the improved mode of performing them. The velocity of the sap varies, however, very considerably, owing to a variety of causes: the nature of the plant, the degree of temperature, and, above all, the quantity of solar light; which last is absolutely required to enable the leaves to evaporate the superfluous water.

During the spring there is a more than usual absorption of sap, for the purpose of nourishing the young buds which are to be developed; and it is very worthy of remark, that the sap which feeds these buds passes through different channels from that which serves to nourish the plant generally. Instead of rising through the young wood, it ascends nearly in the centre of the stem, in the parts contiguous to the medullary channel, and is thence transmitted, by what means is not yet ascertained, through the several layers of wood to the buds.

EMILY.

But the sap that nourishes the buds which first shoot in the spring, cannot have been passed through the leaves, and undergone that change which you say is necessary to convert it into appropriate food. Can it feed the buds in the crude state in which it rises in the stem?

MRS. B.

There is great reason to suppose that it is in some measure elaborated during its passage from the roots to the buds; probably by the organs which it traverses in passing laterally from the centre to the circumference of the stem or branch; but it is a point very difficult to ascertain, owing to the extreme minuteness of these organs: it is, however, a very reasonable inference, since the sap, when it reaches the buds, is in a state ready to be assimilated to their substance.

Part of the sap, which rises in the month of August, in all probability follows the same course, being destined to nourish the new buds which shoot at that season; but it is less abundant than that of March, having fewer buds to bring forth.

EMILY.

How much this sap, destined for the nourishment of the young buds, resembles the milk of animals, a provision which Nature has made for a similar purpose, and which is secreted from the common stock of nourishment only when there are young to feed on it.

And pray what is the cause that produces the rising of the sap in spring?

MRS. B.

Heat is the circumstance most favourable to the absorption of this *nursling* sap, as it is heat which first expands the buds, and makes room for it. An experiment has been made by placing two pieces of vine in two similar vases of water, and then introducing the stem and branches of one of them, through a hole in the wall, into a hothouse: the buds of this plant were rapidly developed, and the water in the vase as rapidly absorbed; whilst the buds of the other plant made only the usual progress, and the water in the vase diminished in the same slow proportion.

If plants are pruned in the spring, the sap will rush out often with violence: in vineyards, this flowing of the sap, when plants are cut, is called the tears of the vine. Mr. Hales made an experiment by cutting off the upper end of the branch of a vine, and enclosing the wounded extremity of the lower part (which remained on the stem) in a tube; the sap flowed from it with such violence, and in such abundance, as to rise to the height of forty-three feet in the tube, thus sustaining the weight of one atmosphere and a half.

CAROLINE.

What a prodigious force! Of course, if you make an incision into the stem of a tree in the spring, the sap will flow out.

MRS. B.

No, not at least with violence, for the spring sap rises with force, only in shoots of one year's growth, and will consequently flow with velocity from none but these. In making the incision, you must penetrate to the centre in order to reach the full channel of the spring sap, and the instant your instrument reaches the pith, you will hear the sap gush, and see it follow the instrument as you draw it out.

We must not, however, bestow the whole of our attention on this nursling sap, but return to that which rises through the alburnum to feed the mature plant. This sap reaches the leaves without having undergone any change; but as soon as it arrives there, a considerable portion of its water exhales by the *stomas*, leaving the nutritive particles which it held in solution deposited in the leaf.

EMILY.

And pray, what is the proportion of the quantity of water evaporated to the whole quantity absorbed by the roots?

MRS. B.

It varies exceedingly, according as circumstances are more or less favourable to evaporation. A plant can evaporate only in proportion to its absorption: the quantity, therefore, depends not only on the abundance or deficiency supplied from the soil, but also on the number of ramifications of the roots; that is to say, of mouths to suck up water. On the other hand, these mouths, however numerous and abundantly supplied, can continue to receive water only in proportion as the exhalation by the leaves carries off what has already been taken in, so as to make room for more. Thus while water enters at one extremity of the plant, it must find its way out at the other.

EMILY.

Were you to pour water into a tube closed at the opposite end, it would soon be filled, and, though you continued to pour, it would receive no more, and the water would flow over; but if you opened the closed end, you might pour in at one end as fast as it flowed out at the other. But what is it that promotes the flowing out, or, in other words, the evaporation, of the water by the leaves?

MRS. B.

The most essential circumstance is *light.*

CAROLINE.

You surprise me: I should have thought that heat would have been more necessary than light to produce evaporation.

MRS. B.

Heat augments it mechanically: but without light no exhalation from the leaves will take place; and it will even be inconsiderable, unless the sun's rays fall upon the plant.

CAROLINE.

Is it not very singular that light should be most favourable to the ascension of the sap which passes through the alburnum, whilst heat is most congenial to that which rises through the centre of the stem? What is the reason of this difference? For both saps, I conclude, must be of the same nature, since the spongioles cannot choose, but must suck up whatever is sufficiently fluid to enter their pores?

MRS. B.

Being derived from the same source, they were, no doubt, originally of the same nature; but when separated into different channels a difference arises: the nursling sap, we have concluded, undergoes a preparation in its passage towards the buds, and their expansion, produced by *heat*, is alone required to call it up.

While the sap which passes through the alburnum must not only throw off a considerable quantity of its water by the leaves, but also undergo a chemical change, for both of which processes you will find that the aid of the solar rays is absolutely required.

Let us first consider the simple evaporation by the leaves. The quantity of water exhaled by plants, is to that which they absorb generally in the proportion of two to three; one third only, therefore, remains in the plant, and becomes a part of its substance; the rest may be considered simply as a vehicle which Nature had employed to convey a due quantity of nourishment into the plant, and which, after having deposited its cargo, disappears.

EMILY.

Is the water then which is evaporated perfectly pure?

MRS. B.

It does not contain above a ten-millionth part of the foreign matter which it held in solution when absorbed, — a very trifling per centage for the expenses of freight.

This exhalation is not visible, because the water is so minutely divided as to be dissolved by the atmosphere as soon as it comes in contact with it.

CAROLINE.

It may then be compared to our insensible perspiration.

MRS. B.

True; and it is called by many botanists the perspiration of plants, and it sometimes happens (as is the case also with animal perspiration) that it becomes sensible. This occurs only in plants whose leaves have simple ribs uniting at a point at the extremity of the leaf. The sap is accumulated by the absorption of the roots during the night, and that portion of it which is destined to be evaporated flows towards this sole aperture, and may be seen there in the form of a minute drop, if observed before sunrise, for it is reduced to vapour by the first solar rays; the subsequent evaporation being equal to the absorption, no accumulation takes place, and no fluid is perceptible. This effect may be seen on the leaves of corn, which, with all the gramineous family, have simple ribs.

CAROLINE.

Plants, then, must increase in weight during the night, since they absorb by the roots without exhaling by the stomas?

MRS. B.

They do so; and whenever, through any accidental cause, the stomas are obstructed or diseased, the plant be comes dropsical, from the accumulation of the water it has taken in and cannot discharge. Plants growing in vases in a room are very subject to this malady, owing to their not having sufficient light to evaporate freely.

EMILY.

Yet if you expose a nosegay in a room to the sun's rays, it withers.

MRS. B.

Because the sun produces a degree of evaporation which the poor mutilated flowers are unable to support; for though the stalks may be immersed in water, the organs of absorption are wanting, and the quantity of water they suck up is quite inadequate to the evaporation. Since, therefore, you have deprived them of the power of absorption, you must diminish, at least, that of exhalation, and, by keeping them in some degree of obscurity, endeavour to preserve the sap which they already contain.

EMILY.

I should be curious to know what quantity of water a plant exhales in a day.

MRS. B.

It has been ascertained by Mr. Hales, that a full-blown helianthus, or sunflower, placed under advantageous circumstances in regard to light and temperature, evaporated twenty ounces of water per day, which is seventeen times more than that evaporated by a man, supposing their surfaces equal. This experiment was made by weighing, first, the water in the vase in which the sunflower was placed, then the plant itself; and, after due time being given to the experiment, the water and the plant were again weighed. The plant had absorbed as much water as the vase had lost; but it was not found to have increased in weight so much as the water in the vase had diminished by twenty ounces, which affords a conclusive proof that these twenty ounces had been evaporated. Of course, suitable precautions had been taken in order to prevent any immediate evaporation from the water contained in the vase. Fleshy fruits, such as apples, plums, peaches, &c. have few or no pores: they therefore retain the moisture they receive from the sap, which enables them to remain long on the tree, after coming to a state of maturity, without drying up and withering. Whilst dry fruits, such as peas or beans, wither in consequence of the number of their pores by which they exhale moisture. There is the same difference between thick fleshy leaves, such as those of the cactus and other succulent plants, and dry leaves, such as those of the pine and the fir, which are at the opposite extremity of the scale; common leaves bear a medium, between the two, but, in the same space in which a common leaf contains six or seven stomas, the leaf of a pine has sixty or seventy.

EMILY.

Aquatic plants which live wholly under water, you told us, were not provided with stomas; but now that I comprehend the nature of their functions, I do not understand why the plants should not derive benefit from them: for while the roots absorbed the water holding ingredients in solution, the stomas would evaporate it in a pure state, leaving all its riches behind.

MRS. B.

The plant has not power to exhale water into water: it requires the assistance of the air to dissolve it and carry it off. Those aquatic plants which rise to the surface, are abundantly furnished with stomas to disburden themselves of their excessive supply of water.

Let us now turn our attention to the nature of the sap which remains in the leaf, after having disengaged its superfluous moisture. It consists of about one-third of the water originally absorbed by the roots, but augmented and enriched by the acquisition of all the nutritive particles which the evaporated water has deposited.

CAROLINE.

In this state it is certainly better calculated to nourish the plant; and from this ample store I suppose the various organs select and assimilate the food they each require.

MRS. B.

It is true that every organ performs a chemical change on that part of the sap which it assimilates to its own substance; but the sap previously undergoes a general change, in some measure analogous to that which the blood undergoes in the lungs, to prepare it for assimilation. This operation is also performed in the leaves, which may be considered as the laboratory in which the sap is submitted to a regular chemical process.

EMILY.

This, indeed, bears a very striking resemblance to the chyle, which is the sap of animals, and which is converted into blood, fitted to go through the general circulation, and nourish the several parts of the body.*

MRS. B.

The analogy is perhaps even stronger than you imagine; for this process, which in animals is performed by means of breathing atmospherical air, in vegetables is performed by the same air acting on the sap when it comes in contact with it at the stomas: the leaves may therefore be considered as the lungs or organs of respiration of plants.

EMILY.

How curious! their stomas then are so many little breathing mouths. And does the oxygen of the atmospherical air carry off carbon from the sap, as it does from the chyle?

MRS. B.

On the contrary, carbon or charcoal is the principal ingredient of wood and of all vegetable matters: the object to be aimed at is therefore to increase, instead of to diminish, the quantity contained in the sap; and the chemical process to which this fluid is submitted in the leaves, though analogous to that performed by the lungs, so far as it prepares the sap for being assimilated to the plant, is rather opposed to it, so far as regards its chemical results.

We animals, the most favoured part of the

* See Conversations on Chemistry, vol. ii. p. 284.

creation, endowed with the faculty of locomotion, require to be of a lighter structure than our tough woody neighbours who are attached to the soil; and, in order to move about with facility, it is necessary for us to disencumber ourselves of part of the carbon we consume in feeding on vegetables; and a man you know, exhales in breathing no less than 11 oz. of charcoal per day; whilst the vegetable kingdom, far from suffering from excess of carbon, requires its store to be augmented.

EMILY.

Ah! this is what I have heard spoken of as one of the most beautiful dispensations of Providence: the vegetable creation purifies the atmosphere, by absorbing the carbon with which it has been contaminated by the breath of animals.

MRS. B.

Just so; but let us examine these wonders a little more narrowly, and trace the steps by which they were brought to light.

Mr. Sennebier covered a plant which was growing in a pot of earth with a glass bell full of water; and, in the course of a few hours, found a quantity of air within the bell. Whence came this air? Did it proceed from the plant or the water in which it grew? He repeated the experiment with water which had been boiled, for the purpose of depriving it of its air, and in this instance no air was produced in the bell.

CAROLINE.

Of what nature was this air?

MRS. B.

Dr. Priestley ascertained that it consisted of oxygen gas, and conceived that it was produced by the decomposition of the water, which, you know, is composed of oxygen and hydrogen; but then he could not understand why boiled or distilled water, which contains as much oxygen as rain or spring water in their natural state, should not produce this air in the glass bell.

At length Mr. Sennebier, in the prosecution of his experiments, discovered the mysterious origin of this air to be in the carbonic acid, which water, in a natural state, always contains. I trust that you have not so far forgotten your lessons of chemistry, as not to recollect that carbonic acid is composed of oxygen and carbon: the plant absorbs this gaseous acid. It is decomposed in the leaves by the sun's rays: the carbon, which it is essential to the plant to retain, is deposited; and within it the oxygen, which it does not require, flies off by the stomas.

CAROLINE.

Then the little vegetable mouths breathe out pure oxygen, and retain the carbon: this is just the reverse of the operation performed in the lungs.

MRS. B.

You may prove this by a very neat experiment. Place two glass jars over the same water-bath, with a means of communication through the water; fill one of them with carbonic acid, and put a sprig of mint in the other. After some time, a vacuum will be produced in the upper part of the jar of carbonic acid; and a quantity of oxygen gas, corresponding exactly to the quantity of carbonic acid which has disappeared, will be found in the jar containing the sprig of mint. And this cannot be accounted for otherwise than by supposing, that the carbonic acid has been absorbed by the mint, decomposed by its leaves, the carbon retained, and the oxygen evaporated.

M. de Saussure has succeeded in measuring the quantity of carbon which plants thus acquire. He transplanted fourteen periwinkles into vases, seven of which he watered with distilled water, and the remaining seven with water in its natural state. After some days he analysed these plants, and found that the former had not made any acquisition of carbon, whilst the latter had acquired a considerable addition of that substance; their wood being one-sixth heavier than that of the former.

EMILY.

And the periwinkles, which had augmented in weight, had, I suppose, alone given out oxygen by their stomas.

MRS. B.

No doubt; but, in making these experiments, attention must be paid to expose the plants, not only to broad daylight, but, if possible, to the full force of the sun's rays; for the solar light is absolutely necessary to the process of decomposing the carbonic acid. During the night the vegetable laboratory is employed in a very different process; for, in the dark, plants absorb instead of exhaling oxygen.

CAROLINE.

You alarm me, Mrs. B.: this is a sort of Penelope's labour, to destroy during the night the work done in the day. And how is the atmosphere to be purified by these means?

MRS. B.

It is true that this apparent inconsistency requires some explanation. You must observe, that the solid nutritive particles dissolved in the sap, whether of animal or vegetable origin, are combined with a considerable quantity of carbon. The sap therefore contains carbon in two states: in the one gaseous, combined with oxygen, and mixed with the water of the sap; in the other combined with different solid ingredients, but dissolved in the water of the sap. The carbonic acid, we have already observed, is decomposed in the leaves, the carbon is retained, and the oxygen thrown off; but what becomes of the carbon contained in the animal and vegetable matter which the sap holds in solution?

CAROLINE.

I suppose it is assimilated to the substance of the plant, together with the other nutritive ingredients which the sap holds in solution.

MRS. B.

No, that cannot be; for, in order to render carbon fit to be assimilated, it appears to be necessary that it should previously be combined with oxygen, and afterwards separated from it.

CAROLINE.

Is there not something paradoxical in this? How can it be necessary that the carbon should be combined with oxygen, merely for the purpose of being separated from it?

MRS. B.

It is very possible that this chemical process may produce a more minute subdivision of the particles than any mechanical operation could effect, and thus prepare it for being assimilated to the plant.

CAROLINE.

Oh, then, now I guess it. During the night the leaves absorb oxygen, to combine with this carbon, and convert it into carbonic acid; and, when the sun rises, this acid is decomposed, the carbon deposited in a state fit to be assimilated, and the oxygen escapes.

MRS. B.

You are right; and as the decomposition of the carbonic acid, which existed in that state in the sap, takes place at the same time, these two operations, being both similar and simultaneous, are confounded together. But, so far as regards the purification of the atmosphere, it is necessary to distinguish them; for, in the first instance, the oxygen exhaled is a mere restoration to the atmosphere of oxygen which had been taken from it during the night; whilst, in the latter, the oxygen evolved, being drawn from the soil with the sap, is so much clear gain to the atmosphere.

CAROLINE.

Well, I breathe freely again, since I know that the atmosphere positively acquires oxygen from the vegetable kingdom. The portion absorbed during the night, I suppose, is but inconsiderable.

MRS. B.

Not so trifling as you seem to imagine; but, since the whole quantity is restored to the atmosphere during the day, you need not apprehend any dangerous results from its abundance. The Stapadra, the plant which absorbs least, takes in a quantity nearly equal to its own volume during a night; and the apricot-tree, which is at the other extremity of the scale, absorbs eight times its own volume of oxygen gas.

Succulent plants absorb the least, having the fewest stomas; and, after them, plants which grow in marshes; then evergreens; and, finally, those plants which shed their leaves in autumn absorb the greatest quantity.

EMILY.

It is this, I suppose, which renders it unwholesome to keep plants in a bedchamber?

MRS. B.

It is; but, besides this, I should tell you that those parts of plants which are not green, such as the brown stems and branches of a tree, and also the flowers, absorb oxygen both night and day, but in such very minute quantities, as not sensibly to deteriorate the air.

Let me hear, now, if you can recapitulate the substance of our conversation.

EMILY.

The sap rises in plants through two different channels: that which is destined for the nourishment of buds, in shoots of the first year, passes near the pith, and is thence conveyed by appropriate vessels through the wood to the buds; that which is to feed the plant in general, rises through the alburnum, and is elaborated in the leaves.

MRS. B.

Very well; and in what does this elaboration consist, Caroline?

CAROLINE.

In preparing the sap to be assimilated to the plant by evaporating great part of the water, and increasing the quantity of carbon. The sap contains carbon in two states: first, in that of carbonic acid; secondly, combined in animal and vegetable matter. In the first state the sun's rays decompose the acid, the carbon is deposited, and the oxygen which flies off purifies the atmosphere; in the second state, oxygen is absorbed during the night, and combines with the carbon, with which it forms carbonic acid; this, during the day, is decomposed, and the oxygen restored to the atmosphere. Thus vegetation serves as a counterpoise to the deleterious effect of the respiration of animals.

EMILY.

And should we not add to the contamination of the air by combustion, Mrs. B.? for oxygen is also absorbed in that process.

CAROLINE.

The air of a forest must then be much more wholesome than that of a town, where so many human beings and animals are continually breathing out carbonic acid, and where such numberless combustions are robbing the atmosphere of oxygen.

MRS. B.

No; the constant motion of the air so rapidly

restores the equilibrium, that it has been found, by the most accurate chemical experiments, that the air of a crowded city contained precisely the same quantity of oxygen as the finest air of the country. I do not mean to say that the atmosphere is not more impure and unwholesome in a large town; but this arises from the smoke, and variety of exhalations, which do not circulate so rapidly as the oxygen gas.

The air in a forest is, on the other hand, far from being considered as healthy; the trees impede the circulation more than the houses in a town, the latter being, in some measure, ventilated by the currents of air which flow through the streets.

CAROLINE.

But, then, consider the pure breath of the green leaves in a forest.

MRS. B.

The exhalations arising from the stagnant waters, and the putrefaction of the dead leaves which remain floating in the confined air, more than counterbalance that advantage, and render a dense forest an unwholesome spot to inhabit.

CONVERSATION VI.

ON CAMBIUM, AND THE PECULIAR JUICES OF PLANTS.

MRS. B.

Having traced the sap in its ascent to the extremity of the leaves, and converted it, by the changes it undergoes in that chemical laboratory, into an homogeneous liquid adapted to the nourishment of the plant; we must now, following it in its descent, observe in what manner it performs this office.

The sap, thus changed, assumes the name of Cambium or returning sap, and passes into another system of vessels which convey it downwards, chiefly through the liber, or most internal layer of bark, and a small portion through the alburnum, or young wood; and, as it traverses the several organs, it deposits in each the various matters requisite for their sustenance.

CAROLINE.

Having compared the ascending sap to chyle, Mrs. B., we may find a still greater analogy between the cambium and blood, into which chyle is converted, after having passed through the heart and lungs, and been rendered fit to nourish the animal frame.

MRS. B.

We have already observed, that the chemical changes which take place in the leaves, in order to convert the sap into cambium, are in many respects analogous to those which take place in the heart and lungs, in order to convert the chyle into blood.

EMILY.

True: in both cases the atmosphere is the agent; with this difference, however, that it carries off carbon from the animal system, while it is the means of accumulating carbon in that of vegetables.

CAROLINE.

But if the cambium descends through the liber, how does it find its way in endogenous plants, which have no bark?

MRS. B.

Its passage in monocotyledons has not been well ascertained. It is probable, that the fibres of the wood are the medium through which the sap both ascends and descends; but you may recollect that it is not well ascertained, whether the ascending sap rises through the vascular or cellular system, or through the interstices between them; and the vessels which convey the descending sap being so minute as barely to be discernible by the aid of a microscope, it is impossible to examine them with accuracy. Besides which a still greater difficulty attaches to the investigation of the vessels of endogenous plants: those which grow in our climates being too small to enable them to acquire that degree of vigour which is requisite for a complete developement of their organs.

CAROLINE.

We shall not have the same difficulty to account for the descent of the cambium, as we have had for the ascent of the sap; since it obeys the laws of gravity and descends by its own weight.

MRS. B.

That is a general cause of the descent of cambium, no doubt; but in the weeping willow, and many other trees whose branches are pending, some additional cause is required to produce the motion of the cambium, since it must rise to return into the stem. It has been ascertained that agitation facilitates and accelerates this motion, and consequently increases the vigour of vegetation; for the more rapidly this nutritive fluid circulates through the several organs, the more frequently it will deposit its nutritive particles in them. Mr. Knight has made a variety of interesting experiments on this subject. He confined both the stem and branches of a tree, in such a manner that it could not be moved by the wind. The plant became feeble, and its growth much inferior to that of a similar tree, growing in a natural state. Mr. Knight confined another tree, so that it could be moved only by the north and south winds, and obtained the singular result of an oval stem; the sides accessible to the wind growing more vigorously than those sheltered from its influence. Every species of restraint, and especially such as tends to render plants motionless, impedes their growth. Stakes by which young trees are propped, nailing them to walls or trellises, greenhouses, or confined situations where this salutary air has not free access, check and injure the vigour of vegetation, and render plants diminutive and weakly.

CAROLINE.

But if young trees were unsupported, they would in all probability be blown down by the first violent wind.

MRS. B.

The stake, it is true, is often necessary; but then it must be considered as a necessary evil, and remembered that, whenever it can be avoided, the plant will thrive better without it. It should never be fastened so tightly as to prevent all motion, for the exercise which the wind gives to young trees is no less salutary than that which a mother gives to her infant; but it is true that the wind is often a rough nurse, over whom it is prudent to keep a watchful eye.

CAROLINE.

Then nailing fruit-trees against walls must be prejudicial to their growth?

MRS. B.

No doubt; but the advantages resulting from the shelter afforded by walls and the heat reflected by them, more than compensate for the bad effects of confinement — for such fruits, at least, as require a higher temperature to ripen them than is to be met with in our climate; but, when the temperature is genial to the plant, standard trees, growing freely in their natural state, produce the finest fruits. Greenhouses and hothouses, however confined, are asylums necessary in winter for the culture of plants of a warmer climate; for though gentle breezes may be beneficial to fan delicate plants, we must shelter them from the inclemency of boisterous winds.

The cambium, we have observed, descends almost wholly through the liber, or most internal and youngest layer of the bark; if, therefore, you cut a ring completely through the bark, this fluid will be arrested in its course, and, accumulating around the upper edge of the intersected bark, cause an annular protuberance. The descent of the cambium thus being obstructed, it will accumulate in that part of the tree above the intersection, afford it a superabundance of nourishment, creating a proportional vigour of vegetation, and a corresponding excellence and profusion of produce.

EMILY.

Would it not then be a good mode of improving the produce of fruit-trees?

MRS. B.

This operation, which is called ringing, has been tried on the branches of fruit-trees, and, I understand, often with success; but I should conceive that the tree must be ultimately injured by the operation; for, if you confine to one part of a plant the food which was destined for the nourishment of the whole, you interfere with the order of that wisest and best of agriculturists—Nature. When interrupted, however, in her original course, she is fertile in expedients to accomplish by collateral means her destined purposes. I observed that some small portion of the cambium descended through the alburnum, which is contiguous to the liber. When the annular section is made on a branch, a much more considerable quantity forces its passage through this channel, and, by affording the young wood an unusual supply of nourishment, renders it harder and heavier below than above the intersection.

CAROLINE.

But if the vegetation of the tree above the annular section is improved, and the wood beneath it better nourished, what part of the plant suffers by this operation?

MRS. B.

Not any part during the season the annular section is made; the evil is reserved for a later period, as I shall explain to you.

The cambium being thus diverted from its course, the greater part being forcibly detained above the annular section, and what little makes its escape descending through another channel, the bark is wholly deprived of its natural sustenance; the consequence of which is, that the new layers, both of alburnum and of liber, which should be annually produced by the descent of the cambium, are not formed.

The following season, therefore, the sap, instead of rising through the soft and tender vessels of the newly-formed alburnum, must ascend through the alburnum of the preceding year, under the additional disadvantage of its being unusually hardened by the superabundant quantity of nourishment it has received.

This artificial mode of rendering alburnum hard and mature, suggested the idea of stripping timber-trees of their bark a year or two previous to their being cut down, in order to harden the young external layers of wood, by forcing the whole of the cambium to find a passage through them, and thus convert the alburnum into perfect wood before the natural period. The experiment, when first made, appeared to answer the most sanguine expectations. The cambium, instead of forming new layers of tender wood under shelter of the bark, forced its way through the alburnum, giving it in one season the hardness and consistence of perfect wood. But it was afterwards discovered that the wood thus artificially matured, by being stripped of its bark, and exposed naked and defenceless to the inclemency of the weather, to the encroachment of lichens and creeping plants, and to the attacks of insects and reptiles; receives injuries, which more than counterbalance the advantages of a precocious maturity, and render it totally unfit for building.

Let us now turn our attention to the composition of the cambium, which subsequently becomes a component part of the plant.

We have observed, that about two thirds of the water absorbed by the roots is evaporated by the leaves, one-third only remaining in the plant. This latter portion exists in vegetables in two states: in the one it retains its liquid form; in the other it is decomposed, and enters into a chemical combination with various parts of the plant, so as to be identified with its solid tissue, and in such a manner that desiccation will not make it reappear.

EMILY.

But by a chemical analysis would you not discover it?

MRS. B.

No; because it no longer exists in the form of

water, but is resolved into its constituent elements, oxygen and hydrogen.* I trust you recollect that water is a combination of these two principles.

M. de Saussure weighed the water with which he watered a plant; and, after the most careful investigation by mechanical means, both by preserving the water evaporated, and obtaining that which remained in the plant by desiccation, he could not discover above five-sixths of the water he had given to the plant.

EMILY.

And is it known under what form this sixth portion of water exists, in its new combinations in the plant?

MRS. B.

The oils and resins with which plants abound contain a very large proportion of hydrogen. There are other vegetable substances which abound with oxygen; the water, therefore, which so totally disappears, is doubtless resolved into its two constituents, oxygen and hydrogen, supplying the oils and resins, and the other juices, with such proportions of these elements as they respectively require.

Carbon is obtained by plants from three different sources: from carbonic acid contained in the sap; from animal and vegetable matter dissolved in that fluid; and from the atmosphere.

* See Conversations on Chemistry, vol. i. p. 227.

CAROLINE.

Having so many means of procuring carbon, no wonder that plants should lay in so large a stock.

MRS. B.

What part of a plant would you imagine contained the most carbon?

CAROLINE.

I should think the wood, which burns so well because it consists almost wholly of charcoal.

EMILY.

And yet the leaves in which the carbon is deposited, when separated from the oxygen, should contain more of that ingredient than the wood.

CAROLINE.

In that case leaves should be used for fuel in preference to wood.

MRS. B.

Emily is right: the green parts of plants contain the most carbon; and dry leaves make an excellent combustible, but they are too large in volume to form a convenient one.

After the leaves, the bark, especially when green, abounds most with carbon; and, lastly, the wood: the alburnum or white wood contains the least.

CAROLINE.

Green wood then should be most combustible; and yet it is noted for burning badly.

MRS. B.

By green wood is commonly meant wood not sufficiently dried. Whatever quantity of carbon wood contains, it cannot prove a good combustible, unless the water, and other juices injurious to combustion, be first evaporated.

The alburnum, when well dried, burns briskly, because it contains a greater quantity of hydrogen than perfect wood; and it is the combustion of hydrogen, you may recollect, which produces flame: but, owing to its deficiency of carbon, alburnum gives out less heat.

The ascending sap, we have observed, contains also a great variety of earthy and alkaline particles; such as magnesia, lime, silex, potash, and soda. When the evaporation from the leaves takes place, these bodies are deposited, and become constituent parts of the cambium, and are thus conveyed to their several destinations.

The most soluble of the earthy salts, such as lime and magnesia, are naturally most abundant in the sap; and when a plant is burnt, the earths, being incombustible, form the materials which constitute its ashes.

The alkaline salts, potash and soda, being also of a soluble nature, are conveyed in considerable quantities into the sap; when this undergoes evaporation, a large portion of these salts is deposited in the leaves, the rest remains in solution in the cambium, incorporates with the plant, and, after combustion, may be discovered in its ashes.

The silecious particles contained in the plant being, on the other hand, nearly insoluble, enter very little into the composition of the cambium, the greater part remaining in the leaves, where it has been deposited by the evaporating sap; and the fall of the leaf is attributed to the accumulation of this hard earthy matter, which in the course of time clogs and indurates their vessels, so as to render them impervious to the juices requisite to their vegetation. The vessels composing the petiole, in which they are so closely bound together, are more especially liable to suffer from these obstructions: unable any longer to transmit nourishment to the leaf, the petiole dries, withers, and falls off; and the plant is thus disburdened of a useless substance, the accumulation of which would be prejudicial to its growth.

CAROLINE.

It must be confessed, that it is rather a serious remedy to destroy the organ, in order to get rid of the inconvenience with which it is afflicted.

MRS. B.

You must consider, that when Nature constructed these organs in so frail and delicate a manner, it was with the intention that they should be annually renewed: it becomes expedient, therefore, to get rid of the old leaves, in order to make way for the new ones.

Azote, an ingredient chiefly of the animal kingdom, is to be found also, in very small quantities, in vegetables: they obtain it both from the atmosphere, of which it forms the chief constituent part, and from the animal matter of manure.

From the cambium, with all the component parts of which you are now acquainted, a great number of different juices are secreted, such as oil, resins, gum, &c.

CAROLINE.

Just as tears and saliva are, in the animal economy, secreted from the blood.

MRS. B.

There is, it is true, a considerable analogy between the animal and vegetable secretions: they are both drawn from the general nutritive fluid, and each by the means of glands; but, owing to the extreme minuteness of the organs of plants, the vegetable anatomy is very much behind that of the animal kingdom.

EMILY.

In small herbs this must necessarily be the case; but in large forest-trees, I should have supposed that the organs, when fully developed, would have been of greater magnitude than those of animals.

MRS. B.

No; the organs of the oak are not larger than those of the family of mosses. Nor is this singular, if you consider that the leaves and fruit of forest-trees are not, in any respect, proportioned to the size of the plant: — you do not forget the fable of the Pumpkin and the Oak. Every leaf and every flower must contain a system of organs, adapted to the various operations it has to perform, without any reference to the general size of the plant. In the animal economy we are still unable to discover the mode in which the glands elaborate their secretions from the blood: how much less, then, can we expect to penetrate the secret, in a system where the organs themselves are frequently so minute as to elude our sight, the largest not being more than one-twentieth of a line in diameter, and there are some, so small as not to exceed the one-hundred-and-fiftieth part of a line in dimension!

Many ingenious hypotheses have been proposed to account for the secretory action of the glands, both in the animal and vegetable economy, but none have hitherto proved satisfactory. That which appears least objectionable, is the agency of electricity; but it must be owned that the chief argument in favour of this agent is, that we are not yet sufficiently acquainted with its powers to prove the hypothesis which rests upon it to be erroneous.

The chemistry of vegetables, on the other hand, is more advanced than that of the animal kingdom; because the great and mysterious secret, *life* performs a less important part in the vegetable than in the animal economy.

The secretions separated from the cambium by the glands are of two descriptions: the one, destined to remain in the plant, is distinguished by the name of internal secretions, and is elaborated by glands of a cellular form; the other, intended to be conveyed out of the plant as useless or detrimental to it, is for this purpose secreted by glands of a vascular form.

The internal secretions are milk, resins, gums, gum resins, manna, essential oils, and fixed oils.

EMILY.

These are substances with which we are in some measure acquainted, as I believe you explained their chemical composition in our Conversations on Chemistry.

MRS. B.

True; but we are now to examine them rather in a different point of view; and I do not think I then mentioned the secretion of vegetable milk. There are three species of this fluid: the first is that which contains opium, and is extracted from the juice of poppies, lettuces, and some other plants; it is almost always white, but sometimes assumes a reddish or yellowish tint.

The second contains caoutchouc, or elastic gum; which, however different in appearance in the artificial state in which we are acquainted with it, is naturally white and liquid.

It is obtained from several different species of trees in tropical climates, but principally from that which bears the name of Hevea.

When an incision is made in the stem, it flows from the wound, and is collected on the surface of small moulds of clay in the form of bottles, to which, being of a glutinous nature, it adheres. It acquires consistence and blackens in drying, and, when the coating of caoutchouc is of a sufficient thickness, it is beaten to pulverise the mould, which is then shaken out.

The third species of vegetable milk resembles that of the cow, and is the produce of a tree in America, thence called the cow-tree. Mr. Humboldt informs us, that it grows in rocky and unfruitful districts, little calculated for the pasturage of cattle. On the barren side of a rock it rises with coriaceous and dry leaves, which are, during many months of the year, not moistened by a single shower. The branches appear dead and dry; but, when the trunk is pierced, there flows from it a sweet and nourishing milk. At sunrise, this vegetable fountain is most abundant. The natives are seen hastening from all quarters, furnished with large bowls to receive the milk, which grows yellow, and thickens at the surface. This tree is of the family of the Sapoteæ.

CAROLINE.

What a delightful resource it must be to the people of that country, who may repose beneath its shade, while they refresh themselves with the grateful beverage it produces! Does it also yield fruit?

MRS. B.

Every tree yields fruit of some kind, otherwise it could not continue its species; but that of the cow-tree is as yet unknown.

Resins are volatile oils, peculiarly modified by the action of oxygen. Pitch, tar, and turpentine are the most common and the most useful juices of this description; they exude from the pine and fir trees, and are of a thick viscous consistence. Copal, mastic, and frankincense are resins of a more refined nature: the two former, dissolved in oil, form excellent varnishes; and frankincense, you know, is the perfume burnt in all the Catholic churches. The resinous juices flow always in a descending direction: when an incision is made in a tree which yields them, they trickle from the upper edge of the wound.

Gum is a mucilaginous secretion, common to all leguminous plants, and to a great number of trees bearing stone-fruits, such as the cherry, the peach, and the apricot: whenever an accidental fissure is made in the stem, it exudes from it.

Gum-arabic is obtained from the acacia of Arabia by incisions made in the stem.

Gum-tragacanth exudes naturally from the stem of *Astragalus*. This secretion, which accumulates during the night, when little or no evaporation takes place, swells the wood which presses against the bark, and, this dry coating not being susceptible of a similar distention, the gum forces its way through it.

Gum-resins appear to be a mixture of the two vegetable products from which their name is derived, and are common to all umbelliferous plants.

Manna is a saccharine secretion, which abounds in the small-leaved ash of Calabria. It is to be found also, in smaller quantities, in several other trees, such as the larch and the willow.

The essential or volatile oils bear a strong resemblance to resins; they are enclosed in small vesicles, whence they are extracted by pressure. They are imprisoned, in this manner, in the rind of the orange and the bark of the cinnamon tree, in the wood of the sandal-tree, and in a great variety of leaves, such as those of the geranium and the orange, and in flowers of almost every description. In a word, there is scarcely any part of a plant from which essential oil may not be extracted, excepting the seed, from which it is absolutely excluded.

Fixed oils, on the contrary, are almost exclusively contained in the seed, where they constitute the most appropriate nourishment for the embryo plant. There is, I believe, but one exception to this rule, the olive, where the oil abounds not only in the seed, but also in the pulp of the fruit, whence it is expressed to supply our tables, or for the purpose of combustion. Nut-oil, linseed-oil, and all the fixed oils, are not, like the essential oils, enclosed in appropriate vessels, but are lodged in every interstice of the seed. We shall speak more fully of this when we come to examine the organisation of this asylum of the embryo plant.

Let us now proceed to the excretory secretions: they are of much less importance than the preceding, and consist chiefly of vapours and gases exhaled from flowers. Among these we distinguish the vapour of the Fraxinella, which is elaborated by glands sufficiently large to be visible, and is very combustible.

EMILY.

I recollect having seen it burn, by approaching a taper to it; but is not this vapour similar to the exhalations of the odour of plants?

MRS. B.

No; the odours of plants are undoubtedly an excretory secretion, but are not generally of a combustible nature. They are of various descriptions, but it is difficult to determine in what manner to class them, as they affect the olfactory nerves of different people in so different a manner: they have been attempted to be distinguished by the name of aromatic, stimulating, penetrating, sweet. Flowers, with some few exceptions, (such, for instance, as the rose and the violet), exhale their perfume only as long as the plant is living; that which proceeds from the bark, or other parts of the plant, continues to be emitted after death.

Flowers having an ambrosial smell, exhale it only in the evening, after sunset; those which have the odour of musk are always of a yellowish purple colour, and of a dull appearance, corresponding, it is said, with the deleterious nature of their perfume.

The smell of flowers, in general, is considered to be more insalubrious to a person sleeping than awake. Whether it be, that, in the latter case, the animal frame has a more energetic power of resistance to deleterious effects, or from some other cause, is not ascertained.

EMILY.

May not this difference arise from plants giving out oxygen during the day, and absorbing it during the night?

MRS. B.

No; the spasmodic effect produced on the nervous system by the perfume of flowers is quite independent of those operations; and it is sleeping or waking, whether in the daytime or the night, that the difference I mentioned has been observed.

Besides the water which plants exhale from the leaves, there are several peculiar juices elaborated by glands situated on the surface of the leaves. These are furnished with hairs, at either the point or base of which they grow, and may be compared to the hairs which grow at the orifice of the pores of our skin.

EMILY.

How extremely minute must those glands be which can be supported on so slender and frail a stem!

MRS. B.

You may thence form some idea of the diminutive size of the vegetable organs in general. When the secretory gland is situated on the summit of the hair, the liquid it secretes is of an innocent nature; when situated at the base, the secretion is acrid, caustic, and poisonous.

CAROLINE.

This is, no doubt, the case with nettles, which pour their poisonous secretions on the skin, and raise it into blisters.

MRS. B.

The poison must penetrate beneath the cuticle in order to produce this effect; the hair is the instrument which gives the wound, and the poisonous juice is then poured into it.

EMILY.

This is just like the sting of a serpent, who inflicts a wound, and then ejects his poison into it. But what is the reason that nettles do not sting when wetted with rain?

MRS. B.

Because the hair, when softened by moisture, has not sufficient strength to perforate the skin; and, unless a puncture be made, the secretion cannot insinuate itself beneath the skin, and no sting is felt. Stinging plants can also be handled with impunity after death, if dried; for, though in this case the instrument may be capable of wounding, the poisonous juice is no longer fluid, and cannot flow into the puncture.

EMILY.

Then should we not feel the infliction of the wound, although we might escape the smarting of the blisters?

MRS. B.

The instrument is so minute and delicate, that the wound it inflicts would not be felt were the skin not inflamed and blistered by the poison.

The nectar of flowers, the bloom of fruits, and the viscous coating of aquatic plants, which preserve them from the element in which they grow, may all be considered as excretory secretions; but we will postpone their examination till we enter upon the subject of flowers and fruits.

We have now traced the sap, from its first entrance into the roots, throughout the whole frame of the plant; we have examined its component parts, the chemical changes it undergoes in the leaves, its subsequent descent under the form of cambium, and the various peculiar juices which are secreted from this nutritive fluid, as it returns from the extremity of the leaves into the roots.

EMILY.

But is the whole of the sap consumed in the performance of these several operations, and does no part of the cambium return through the roots into the earth?

MRS. B.

It is the opinion of M. De Candolle, founded on experiments he has made on this point, that a small residue of the cambium exudes from the roots into the ground.

In planting, it has long been observed that trees of a different species from those which previously occupied the ground, thrive better than a repetition of those of the same kind: whence it is inferred that the exudation of one species of plant, though it may injure the soil for other individuals of the same species, may possibly afford appropriate nourishment for plants of another description. But this is a question we shall refer to at some future period.

Having now concluded our examination of the structure of a plant, and of the mode in which it is nourished, I shall proceed next to observe in what manner it is affected by the external bodies with which it is in contact, such as light, heat, the atmosphere, the soil, &c. This will enable you to acquire some information respecting the culture of plants; and though I do not aim at making you adepts in agriculture, yet I consider the application of botany to that science as the most useful and the most interesting point of view in which it can be studied.

CAROLINE.

And it must be much more amusing than the common mode, of studying the classification of plants.

MRS. B.

Classification must always be respected in science. It is impossible to acquire any clear ideas in any branch of knowledge without its aid; but it is true that in botany it is sometimes held in too high estimation. In the eagerness of pursuit, the student is apt to forget that classification is but the means, not the end to be attained.

M. De Candolle's mode of classification is more simple than that of any preceding botanist; still you must expect it to afford you instruction rather than amusement. But we shall not treat of it till we have examined the various organs of the flower on which it is founded.

CAROLINE.

And when are we to learn the history of the flower, Mrs. B.? — that part of botany which I once thought comprised the whole of the science?

MRS. B.

The flower is the asylum, in which the seed, destined for the reproduction of the plant, is lodged and cherished. We shall examine the flower and the seed, therefore, in immediate succession, when we enter on the subject of the reproductive powers of plants.

CONVERSATION VII.

ON THE ACTION OF LIGHT AND HEAT ON PLANTS.

MRS. B.

In examining the effect of external bodies on plants, we shall begin with light, which may be considered as acting on them in four different ways. The first rays of the rising sun seem to awaken the vegetable creation from its state of repose.

CAROLINE.

You do not mean to infer that plants sleep during the night, Mrs. B.?

MRS. B.

I doubt whether the term sleep be literally appropriate to that state of relaxation and inaction which appears to afford them repose during that season. The leaves and flowers usually change their position as soon as it grows dark; in many plants the leaves drop; in others they close, as well as the petals of the flowers, and are opened by the first rays of the morning sun. The leaves then recommence their chemical operations, the spongioles draw up a provision for their labours, every function which had ceased or diminished during the night is again renewed, and the whole plant reanimated. It is this effect, produced by light on plants, which I call being awakened.

Secondly, the direct rays of the sun are necessary to enable plants to decompose carbonic acid gas in any sensible quantities. We have already observed, that in this process the oxygen of the carbonic acid is exhaled by the leaves, and the carbon deposited in the plant: now, it is this deposition which produces their green colour. Mr. Sennebier is of opinion, that carbon is not positively black, but of a dark-blue colour. The cellular tissue of plants is of a yellowish white; consequently, when those minute blue particles are lodged within the yellow cells, the combination of the two colours produces green, in which the blue or yellow tint prevails, in proportion as the carbon or cellular tissue predominates.

CAROLINE.

That is very curious, and accounts for the pale delicate tint of the spring verdure, when but a small quantity of carbon has been deposited in the leaves; and the deeper shades which plants acquire in summer and autumn, when they have accumulated a greater stock of carbon.

But what is it that produces the change of colour at the fall of the leaf, and, indeed, often takes place previous to their fall, when some leaves assume a beautiful red or yellow colour?

MRS. B.

Some ingenious experiments have lately been made on this subject by Mr. Macaire. He ascertained that, late in the autumn, when leaves begin to change their colour, they absorb oxygen during the night, but lose the power of giving it out during the day: hence he inferred, that the accumulation of oxygen destroyed the green colour, and produced the various tints which the autumnal leaf assumes.

EMILY.

We know the power that oxygen has in changing the colours of metals: it is not, therefore, surprising that it should produce effects somewhat similar on plants; but if it is oxygen which gives rise to the red and yellow colour of dying leaves, may it not also be the cause of the various hues of living flowers? Mr. Macaire should have prosecuted his researches to have discovered this.

MRS. B.

He did so; and was led by their result to think, that it is to the various quantities and modes of combination of oxygen, that the different colours of flowers is to be attributed.

The third mode by which light acts on plants, is by facilitating, and consequently increasing, their absorbent powers. The more the cause is augmented, the more the effect is increased. Tell me now, what do you suppose would be the result of great intensity of light?

EMILY.

To be enabled to answer your question, it would be necessary to be acquainted with the plants of tropical climates, where light as well as heat is so much stronger than in our latitudes.

CAROLINE.

It would be more easy, in this mountainous country, to study the plants which grow on their summits, where they are so much more exposed to light than in the valleys or the plain; and I recollect observing, that they are generally of a deeper green, which is no doubt owing to the greater deposition of carbon.

EMILY.

I have remarked, also, how much deeper the green colour of vegetation is in Italy than in England.

MRS. B.

In the tropical climates this difference is still more remarkable. But what is very extraordinary, M. Humboldt met with some green plants, growing in complete darkness, at the bottom of one of the mines of Freuberg. The mine abounded with hydrogen. Now, whether this gas be endowed with the power of developing the green colour, or whether it may enable them to decompose carbonic acid without the aid of light, is a problem which we must leave to more able chemists to resolve. What increases the difficulty is, that carburetted hydrogen gas is a poison no less deleterious to plants than it is to animals.

Another effect of intensity of light, is to render plants remarkably firm and hard of texture, owing both to the accumulation of carbon, a body of a compact solid nature, and to the increased vigour of their powers of absorption, which enables them to incorporate a greater quantity of the earthy matter dissolved or floating in the increased quantity of sap they suck up.

CAROLINE.

But, on the other hand, they must contain a greater quantity of liquid, which could produce a contrary effect.

MRS. B.

Recollect that intensity of light increases the power of evaporation, as well as that of absorption; the plant, therefore, retains no superabundance of liquid, although it acquires more of the solid particles which it held in solution. The compactness and hardness of plants exposed to an excess of light, offer some impediment to their growth; their vessels, deficient in elasticity and flexibility, are not so susceptible of being elongated by the fluids which circulate within them. Mountainous plants are therefore more diminutive in size than those of the plain.

EMILY.

But this is far from being the consequence of exposure to light in hot climates, where the vigour of vegetation greatly surpasses that of our more moderate temperature, and where plants are in general of much larger dimensions than with us.

MRS. B.

It is to the intensity of heat, rather than of light, that they owe their superiority. When these two causes act simultaneously on plants, they concur in promoting the vigour of vegetation. The intensity of heat tends to distend the vessels indurated by the deposition of carbon, and to accelerate and give increased impetus to the motion of the sap; which, flowing profusely through these firm yet still elastic vessels, produces a force of vegetation, and a magnitude of dimensions, unknown in our less genial climes.

The mountain plant, on the contrary, is peculiarly exposed to cold; the chilled and languid sap traverses with difficulty the indurated vessels, the circulation of all the juices is checked, and the vigour of vegetation proportionally diminished.

CAROLINE.

And yet the flowers on mountainous plants appear to me unusually large.

MRS. B.

It is an observation which has been frequently made; but I am inclined to think it is an illusion, produced by the comparative diminutiveness of the stem and branches.

There is a third cause of the greater hardness of plants exposed to intensity of light. By assimilating a more considerable quantity of carbon, the plant at the same time decomposes and incorporates a greater quantity of water. It is not known how this operation is performed; but the water, no longer in a liquid form, augments instead of diminishing the solidity of the plant.

CAROLINE.

I should be curious to examine also, the other side of the question; that is to say, the effects of a deficiency of light, such as occurs with plants cultivated within doors, in confined situations, in the shade of forests, &c.

MRS. B.

In the first place, they are pale from not having a sufficiency of carbon to develope their colour. In consequence of this deficiency of carbon, their fibres, being soft and feeble, are èasily stretched out, and grow to a great length; as you may possibly have seen potatoes, when sprouting in a dark cellar, shoot out weak slender branches, six or eight feet in length.

A deficiency of light diminishes the power of evaporation still more than that of absorption, so that the plant retains an excess of liquid, and becomes literally dropsical. This state of saturation diminishes both their smell and their savour. Advantage is taken of this circumstance to soften the flavour of vegetables when too strong; that of celery, for example, is tempered by burying the stem in the ground, and sheltering it from the light, the leaves alone are suffered to appear above ground. In these the green colour is developed, while the stem remains perfectly white.

CAROLINE.

But since it is the leaves which occasion the deposition of carbon, I should think the purpose would be more effectually accomplished by covering *them* up with earth also, or by stripping them off the stem. And by depriving the plant of its means of evaporation, you would also increase the retention of sap, and render the plant more tender and less-strongly flavoured.

MRS. B.

But such extreme measures would check vegetation, and render the plant diseased, if not actually destroy it.

It is with the view of making lettuces white and tender that they are tied up, so that the external leaves are alone exposed to the light.

CAROLINE.

It is not necessary for the gardener to perform

such an operation on cabbages; for Nature takes this precaution so completely, that it is the external leaves only which develope any colour, the heart being quite white and tender, from a deficiency of carbon and a superabundance of water. Endive is also artificially whitened, and its flavour softened, by being covered with tiles; the green leaves of endive, which have not been thus sheltered from the light, are very unpleasant to the palate from their bitterness. But vegetables thus whitened, though tender, are generally crisp, not soft.

MRS. B.

That crispness would have been converted into a hardness approaching to woody fibre, if the plant had not been sheltered from the light. The crispness is very agreeable to the palate in lettuces and endive when eaten raw, and it becomes perfectly soft by cooking; whilst those parts of a vegetable in which the woody fibre begins to be developed, by the deposition of carbon, after cooking remain tough and stringy. You must have noticed this difference in every dish of cardoons and of celery which is served at table: those parts from which the light has been completely excluded are quite soft and tender, whilst those ribs which have been partially exposed to it, are more indurated and fibrous.

It was supposed by the ancients, and, indeed, taught by their great naturalist, Aristotle, that the verdure of plants was developed by the atmosphere, and that it was the exclusion of air which prevented the roots from assuming that colour.

EMILY.

What would be the consequence of depriving a plant entirely of light?

MRS. B.

The leaves would become dropsical and fall off: fresh leaves would, indeed, sprout; but these, having no power to decompose carbonic acid, would be completely etiolated. Deprived of carbon, and in a great measure of the earthy matter deposited by evaporation, the new shoots would be soft and feeble, but considerably elongated by the absorption of fluid which they have not power to throw off.

Mr. Bonnet of Geneva cultivated some peas in a cellar totally dark, and they were completely etiolated.

EMILY.

Pray, can you tell me why plants turn their leaves and flowers, and even stretch out their branches towards that side on which the light predominates?

MRS. B.

Some have imagined that this preference resulted from a sort of instinct; others have gone so far as to discover in it an indication of sensibility.

EMILY.

I begin to take such an interest in plants, since you have made me better acquainted with them, Mrs. B., that I should be delighted to find they were raised something above the mere passive mechanical beings, in which Nature carries on her chemical and physical processes without their interference.

MRS. B.

In all organised beings, life plays so considerable a part, as effectually to distinguish them from mere brute matter; but, in regard to the instinct and sensibility of the vegetable creation, I fear we must abandon the subject to poets, who have often treated it with much beauty.

The inclination of plants towards the light, we sober-minded botanists account for in a far less romantic manner.

I must relate to you an experiment made by Mr. Texier, which shows that plants, far from being endowed with feeling, are mechanically compelled, by their kind parent Nature, to turn towards that quarter which is most conducive to their well-being. Mr. Texier placed a plant in a dark cellar, where it was supplied by openings on the one side with light, and on the other with air; and the plant was left at liberty to give the preference to either. In a short time the point was decided: both leaves and branches extended themselves towards the light. This partiality may, perhaps, be thus explained. Those parts of a plant most exposed to the light becoming harder and less susceptible of elongation than the parts more in the shade, the two sides being unequal, the one is obliged to yield to the other; the soft, yielding, elongated side to that which is harder and more contracted. If one side of a branch is more elongated it will take a curved form, as may be seen in the figure (Plate IV. fig. 5.); where the dark line represents the accumulation of carbon and contraction of growth, and the fine line, the softer texture and greater elongation; and thus you see that the plant mechanically assumes a position in which it may receive the greatest benefit from an element so essential to its welfare.

CAROLINE.

This is extremely curious; and it accounts for the tendency of plants towards the light in a manner so simple and clear, that, mechanical as it is, I cannot doubt its correctness.

MRS. B.

It has been suggested, that advantage might be taken of this natural mode of curving wood for the construction of ships, the ribs of which are bent by artificial means.

Let us now proceed to examine the effect of temperature on plants.

Heat excites and accelerates the circulation of the juices of plants, as it does those of the animal frame; but this effect varies in different species of plants, and even in different individuals of the same kind. It is the accelerated motion of the sap, during the warmth of spring, which determines the period of the budding of the plant; and, as the temperature of the season increases, produces a greater absorption by the roots, of evaporation by the leaves, developes the blossom, and, finally, ripens the fruit.

The action of heat in these operations is not merely mechanical, but produces effects on the living plant very analogous to those which it does on animals.

When, on the contrary, the temperature of the soil is as low as the freezing point, the spongioles finding no fluid to imbibe, the plant, deprived of sustenance, languishes; and, should this privation continue any length of time, it perishes of famine.

EMILY.

And if the heat be so intense as to evaporate all the water of the soil, the plant will be equally deprived of nourishment.

MRS. B.

The effect is similar, though produced by so opposite a cause. The plant will in the last case, however, perish sooner; for, besides being deprived of sustenance, it will evaporate its own moisture. But let us first enquire into the effects of a low temperature on plants. Water may freeze within the plant, but it is less liable to congelation after, than before being absorbed by the roots; not only because it is better sheltered from the external cold, but because the motion of the sap is unfavourable to congelation.

If the stem of a tree freezes, the elasticity of the vessels requisite for the circulation of the juices is destroyed, and the plant perishes both from cold and hunger. But if merely the leaves and buds be frozen, they alone are destroyed; and the sap, which the stem continues to transmit to the branches, enables them to sprout out new buds and fresh leaves.

CAROLINE.

So, a man would still live, were his nose or fingers frozen. The analogy, however, will go no further, for he has not the advantage of sprouting out new ones.

And, on the other hand, what degree of heat will plants support?

MRS. B.

It varies extremely, depending on a variety of circumstances. The *Vitex agnus castus* has been known to strike root in water at the temperature of 170 Fahrenheit.

When one of the hothouses of the Botanical Garden of Paris was burnt, all the plants within it perished excepting the flax of New Zealand, which resisted the temperature of a general conflagration that consumed even its leaves.

The temperature of some few plants experiences an elevation at the moment of flowering: that of the *Arum maculatum* rises from fourteen to twenty-one degrees, when its blossom expands between three and five o'clock.

EMILY.

Do plants, like unorganised bodies, partake of the temperature of the atmosphere in which they grow? I should have imagined that they must be warmer; for, since some portion of the water which a plant absorbs remains incorporated in it under a solid form, in changing its state from fluid to solid, it must give out its latent heat, which would raise the temperature of the plant.

CAROLINE.

But you forget, Emily, that two thirds of the water which the plant absorbs is evaporated, and, by changing its form from liquid to vapour, cold must be produced. Now, as the quantity which assumes the form of vapour is much greater than that which becomes solidified, the temperature of the plant should ultimately be lowered beneath that of the atmosphere in which it lives.

MRS. B.

Each of your opinions have been sanctioned by different naturalists; but, however exact these calculations may be in the laboratory, they can give us but very little insight into the chemistry of vegetation, where that mysterious principle *life* performs so essential a part.

EMILY.

Yet, is it not easy to ascertain by the thermometer whether plants differ in temperature from the atmosphere?

MRS. B.

It appears by that test, that the trunk of a tree is colder in summer, and warmer in winter, than the air by which it is surrounded; and the larger the stem, the more this difference is manifest. But the reason of this is very simple: the roots draw their nourishment from a depth of soil, well sheltered from both extremes of heat and of cold: the water they absorb remains throughout the year of a moderate temperature; and the stem, which serves as a channel to transmit this water to the branches, naturally acquires the temperature of the fluid it conveys. The sap thus tends to cool and refresh the plant during the heat of summer, and to cherish and preserve it from being injured by the severity of winter; but should this severity be so intense or of such long duration as to penetrate into the deep recesses of the soil, whence the sap is drawn, the temperature of the tree will gradually descend to that of the atmosphere.

CAROLINE.

The stem has still another defence from the cold, in the several layers of bark; which we may, I suppose, consider as so many warm coats to preserve the internal temperature of the tree?

MRS. B.

Certainly. The bark is a bad conductor of heat, and, like flannel-clothing, serves equally to keep in warmth during winter, and to exclude heat in the summer.

Notwithstanding all these precautions, which Nature has so wisely taken to preserve the equable temperature of plants, the water within the stem is sometimes frozen; and, as water when converted into ice occupies a larger space than in its fluid state, it bursts the vessels in which it is contained, and injures, if it does not destroy, the texture of the plant.

EMILY.

But should this occur after a dry season, so that the cells which contain the water be not filled, there will be room for its expansion in freezing.

MRS. B.

True; freezing may then take place without considerable injury to the plant: it is only when the parts become disorganised, from the fracture of the vessels of the cellular system, that the plant itself is said to be frozen. The more water, therefore, plants contain, the more liable they are to injury from frost; and, accordingly, we find that aqueous plants are those, most easily frozen. What parts of a plant should you suppose most liable to be attacked by the frost?

CAROLINE.

Those which contain most water: the leaves, where that liquid is conveyed to be evaporated; and the buds, where, during spring, the sap is brought in such abundance for their nourishment. Besides, the leaves and the buds are most exposed to the air, while the stem and branches are well defended from its inclemency by their warm clothing, the bark.

MRS. B.

You have judged rightly. If the frost be so inveterate as to attack the stem, it is the alburnum, as being the most moist and tender, which first suffers; afterwards the liber, the internal coating of bark. If this freezes, death must ensue, as the vessels which convey the cambium are lodged in this coating. The external layer of bark is the driest of any part of the plant, being constantly subjected to the inclemencies of the season. It is often injured, and in the lapse of time destroyed, and peels off; but it is never frozen.

EMILY.

Plants must be more liable to freeze in the

spring than in the autumn, at an equal temperature, because they contain more water in the former than in the latter season.

MRS. B.

True. In autumn the absorption of sap diminishes, in the spring it increases, both in quantity and celerity of motion, in order to provide for the budding of the plant; and if in this season a frost should prevail, there is great danger of the plant falling a sacrifice to it.

Plants of different species vary much in their power of resisting cold. Oaks do not freeze at 56° below the freezing point of Fahrenheit; beech will support 79°, an intensity of cold which congeals mercury. Mr. De Candolle found snowdrops in blossom on Mount Saleve beneath the ice; and Captain Parry, in his Polar expeditions, discovered many plants, in full leaf and ready to blossom, encased in ice.

It is remarkable that plants, which are the greatest sufferers by extreme cold, are at the same time most liable to injury from intense heat. But this apparent inconsistency admits of an easy solution. We have observed that aqueous plants are easily frozen; they also evaporate abundantly; therefore, when exposed to extreme temperature, whether of heat or cold, they will be either frozen or dried up.

Plants which secrete a viscous juice do not easily freeze: the tenacity of this thick, sticky fluid prevents, or at least impedes, that arrangement among the particles which is necessary to produce congelation. Freezing, you know, is a species of crystallisation; and it is requisite that each particle of liquid should be free to range itself in that order which is essential to the formation of such regular bodies as crystals.

CAROLINE.

The rising sap then must freeze more easily than the descending sap or cambium, as the latter is thicker and more viscous.

MRS. B.

Yes; and there is also another reason why the sap is more liable to freeze than the cambium: the former moves with greater celerity, so that its particles more easily place themselves in the order of crystallisation.

EMILY.

Might not recourse be had to the expedient of stripping the plant of its leaves, in order to diminish the velocity of the rising sap, when in danger of freezing? For by depriving a plant of the organs of evaporation you lessen its power of absorption.

CAROLINE.

But, by preventing the elaboration of the sap in the leaves, you hinder it from acquiring that consistence which enables it to resist congelation;

you render the plant dropsical by the accumulation of water; and aqueous plants, Mrs. B. has told us, are most liable to freeze.

MRS. B.

Emily's expedient has been tried, but not without danger of the consequences you deduce from it.

Fleshy fruits, such as oranges, apples, and pears, requiring a great quantity of sap to supply sustenance, occasion a great absorption by the roots. These plants are, consequently, particularly liable to injury from frost; and, when thus endangered, it is a useful precaution to gather the fruit, in order to secure the tree. In the south of France, the oranges are gathered on the first appearance of a frost; and should this operation not be completed before the frost sets in, it frequently occurs that the side of the tree on which the fruit remains is attacked by the frost, whilst that on which it has been gathered escapes uninjured.

CAROLINE.

Then, in cases of such urgency, they should begin by gathering the fruit on the north side of the tree, as being most exposed to the cold.

EMILY.

Some fruits, like the peach, are coated with a soft down; which, I suppose, answers the purpose of warm clothing?

MRS. B.

Yes; it is perhaps, even a better preservative from the cold than the coàtings of the epidermis. This soft down encloses and confines the particles of air on the surface of the fruit on which it grows. Air, you may recollect, is a very bad conductor of heat, and especially air in a tranquil state; that which is imprisoned by the down affords, therefore, the most useful shelter to the plant.

EMILY.

I have often experienced the advantage of a precaution of this nature, by holding up my fur-tippet before my mouth when encountering a sharp frosty wind; the air, held captive by these slender threads, reposes tranquilly in its downy prison, and becomes mild and genial to breathe.

MRS. B.

You must observe, also, that, during its captivity, it is tempered by the warmth of the breath you expire, before being inhaled by the lungs; so that, in fact, you breath a tepid, instead of a frosty air.

A layer of air is also retained captive between the epidermis and the bark, which is, perhaps, a still better preservative to the plant than the bark itself: it is a delicate under-garment, which the stem wears beneath its more cumbrous clothing of epidermis.

CAROLINE.

The epidermis was itself, I thought, a delicate covering to the internal layers of bark.

MRS. B.

That depends upon the nature of the plant, and the part to which it belongs. The epidermis of the leaves and buds is delicate, but that of the stem and branches of a venerable oak is of a very different description: the bark in general which covers the trunk consists almost wholly of carbon; which, being a very bad conductor of heat, answers the double purpose of confining the internal heat in winter, and excluding the external heat of summer.

The epidermis itself is sometimes single, sometimes double or triple, but more commonly consists of a number of layers. So many as one hundred and fifty have been counted in the epidermis of a tropical plant; and so great a number still remained that the calculation was abandoned from the difficulty of completing it.

At our next interview we shall examine, how far plants will admit of being naturalised to a climate differing in temperature from that in which they are indigenous; and what are the precautions necessary to be taken in transplanting them to a foreign country.

CONVERSATION VIII.

ON THE NATURALISATION OF PLANTS.

MRS. B.

In estimating the effect of diversity of climate on plants, the point most important to be considered is the difference of temperature. The nature of the soil, the air, water, and light, are circumstances comparatively trifling, compared with the abundance or deficiency of heat.

EMILY.

I should have imagined that the quality of the soil and the quantity of water would have been of still greater consequence than the temperature.

MRS. B.

When we wish to naturalise a foreign plant, art may do much in rendering the soil analogous to that in which it originally grew, in affording it a due quantity of water, in sheltering it from, or exposing it to the light. The nature of the air varies very little in any latitude, but its temperature most remarkably; and over this art has little or no control.

CAROLINE.

You forget our hothouses, Mrs. B., where we produce whatever temperature we choose.

MRS. B.

True; but the plant cultivated, or I should rather say forced, in such an artificial atmosphere remains a foreigner, and does not become a naturalised subject of the vegetable realm.

If you compare the mean temperatures of different countries, you will be surprised to find how much more nearly they approximate to equality than you would at first imagine. For instance, those of England and of Switzerland do not vary above two or three degrees; yet they frequently will not admit of the cultivation of the same plants. In Switzerland it is hotter in summer than in England, owing to the latitude; whilst its local elevation, and the vicinity of mountains covered with snow, renders it colder in winter. The more equable temperature of England, throughout the year, enables every species of laurel, and even Rhododendrons, to support the winter with impunity. In Switzerland the common laurel, if it escape being frozen, suffers so much as greatly to injure its vigour and its beauty: the contrast of a strong vegetation in summer, suddenly checked by the severity of winter, ill accords with its nature. The laurustinus and the Portugal laurel are unknown in Switzerland, except as greenhouse plants; whilst, on the other hand, the fruit of the vine, which we can but imperfectly ripen in England, in Switzerland affords a luxuriant vintage.

EMILY.

I am often inclined to envy them their oleanders and pomegranates, which blossom so beautifully in the open air, and require the shelter of a greenhouse only in winter; while we produce very inferior plants of the same description even in our hothouses.

CAROLINE.

Yet the temperature of Switzerland must be lower than its latitude would indicate, owing to its elevated situation; for being nearly in the centre of Europe, whence almost all the great rivers have their source and flow into the sea, it must be the spot most raised above that level.

MRS. B.

Certainly. In estimating the temperature of a climate, the prevalence of hot or cold winds, such as the *sirocco* in the south of Italy, the *mistral* in the south of France, and the *bise* in the valleys of Switzerland, should be taken into consideration, as well as the locality of the spot; which, independently of its degree of elevation above the level of the sea, is liable to be affected by a variety of circumstances. One of these, which, however remarkable, naturalists have not hitherto been able satisfactorily to account for, is, that in countries of similar elevation and latitude the temperature is always higher in those situated on the western than on the eastern side of a continent. It is warmer, for instance, at Nantes, on the western shore of France, than at Quebec, on the eastern shore of America, both being very nearly of the same latitude. At Quebec, snow-shoes and sledges are in general use during several months of winter, and booths are built upon the frozen river St. Lawrence; whilst at Nantes frost and snow are little known, and of short duration.

EMILY.

But how far can plants accustom themselves to a climate and temperature which is not natural to them?

MRS. B.

It varies extremely, according to the nature of the plant. The horse-chesnut, which is so well naturalised to our northern climates, that it braves even the inclement skies of Sweden, was originally brought from India; where it grew, it is true, on mountains, but of no very considerable elevation. Some plants succeed only partially on being transplanted to a foreign climate. Thus the artificial grasses, such as clover and cinquefoil, thrive very well as grasses: they are cut down when in blossom, the heat of summer being seldom sufficient to ripen their seed, we are under the necessity of importing it from warmer climates, in order to renew them.

EMILY.

I thought that the artificial grasses were cut down at that earlier period of their growth, as being then most tender, and best suited for the nourishment of cattle.

MRS. B.

That is true: but were we able to ripen the seed, we should cultivate a portion for that purpose, instead of annually renewing it from foreign parts.

CAROLINE.

Why have these grasses obtained the name of artificial?

MRS. B.

Because they require continued cultivation from seed. They are not perennial, but must be constantly resown; whilst most of the grasses of our pastures and meadows spread by the root.

There are many plants which will not admit of transplanting to a colder climate: thus the orange and the olive have made no progress northward since the time of the Romans, but are still confined to the same limits.

CAROLINE.

The orange-tree bears our climate under shelter of a greenhouse extremely well.

MRS. B.

There are few plants which cannot be cultivated with some degree of success, by the artificial temperature of a greenhouse or a hothouse; but I am speaking of their being naturalised to a climate, so as to admit of being raised in the open air.

When you make the experiment of introducing a new plant from a warmer climate, you must treat it with great care, and endeavour, by gentle gradations, to wean it from its native country, and accustom it to our more inclement skies. You should begin by placing it in a hothouse; the following year you may try the greenhouse; and, if it does not appear to suffer from this change, you may finally expose it to the open air. While the plant undergoes this species of education, you have the advantage of studying its habits, the nature of the soil most favourable to its growth, the quantity of water it requires, the degree of light to which it should be exposed, the wind which it will support, and a number of minute circumstances, which will indicate the situation and treatment most congenial to it, on transplanting it into the open air. This knowledge of the habits of plants is highly essential to their success, in becoming naturalised to a foreign climate.

EMILY.

Undoubtedly. If it be found that they require much moisture, it will be expedient to plant them in hollows rather than on rising ground. When the plant is tender and delicate, so that light, heat, and shelter become essential to its preservation, you must select a southern aspect, that it may derive every possible advantage from the sun, and be sheltered from the north wind.

MRS. B.

You must also pay attention to plant it rather deep in the soil, in order that its roots may be supplied with water of a moderate temperature, and that the neck or vital part of the plant be sheltered from the inclemency of the weather, by being well covered with earth. Then it should be transplanted in the spring, in order that it may be gradually accustomed to a diminution of temperature, instead of being suddenly exposed to the severity of winter. Besides, if planted in the spring, it will be hardened against the following winter by the deposition of carbon during the summer. It is desirable, also, to plant it in a rich vegetable soil, in order to afford it plenty of nourishment.

CAROLINE.

In Switzerland, they plant rhododendrons and kalmias in pots of remarkably black earth.

MRS. B.

The bog-earth or peat-earth of England, though less rich, is of the same nature, consisting chiefly of vegetable remains. There are whole districts in Belgium of this nutritive vegetable soil, which is converted into nursery gardens for raising kalmias, rhododendrons, and other plants of this description, where they grow and prosper in the open air. In England the rhododendron succeeds perfectly well in our gardens, though the soil is less rich in carbon than the bog-earth of Belgium; but the moisture of our climate is particularly favourable to that, as well as to every species of laurel, and to evergreen plants in general.

In transplanting from a colder climate, very few precautions are required: an elevated situation is desirable, and a sufficiency of water to provide for the more abundant evaporation to which the plant is subjected.

Plants brought from a warmer climate should be watered but moderately, because the power of evaporation is checked by the diminution of temperature. This is particularly to be observed in autumn, when the cold weather first sets in.

EMILY.

And in that season, the direction, I should suppose, would be applicable to plants of every description, none of them being capable of evaporating so much during winter as summer, especially when deprived of their organs of evaporation — the leaves.

MRS. B.

True; but not equally applicable to those which preserve them. Such greenhouse plants, for instance, as geraniums and orange-trees, which retain their organs of evaporation throughout the winter, though that function is more imperfectly performed during this season, will admit of being watered with less parsimony than others.

EMILY.

Do you approve of sheltering delicate plants, by covering them with straw or matting during the winter?

MRS. B.

If the spot in which they grow be elevated, and the soil dry, it may be done with advantage; but in low damp situations such a precaution might occasion the plant to rot, particularly evergreens, as the covering would prevent the evaporation of any superabundant moisture by the leaves. In such cases, it is better to leave the plant exposed, taking care, only, to shake off the snow which, if melted by the sun during the day, runs down the stem and branches, and insinuates itself into any little crevice it may chance to find in its passage, or is absorbed by the buds and tender parts of the plant as it trickles over them. Then at night, if this water freezes it injures, if it does not destroy, the texture of the parts in which it is lodged.

CAROLINE.

It must surely be desirable to plant such trees against a wall in a southern aspect, even under the inconveniences to which such a state of confinement subjects them.

MRS. B.

It is attended with several advantages, which, in the initiation of tropical plants to our climate, often more than compensate the injury resulting from confinement. A southern wall not only affords shelter from the north wind, but becomes a source of heat, by the transmission of the sun's rays to the plant.

EMILY.

The white walls of France and Switzerland must reflect a great deal of heat; but I should not have supposed that our brown brick English walls would have produced much effect, for they must absorb more heat than they reflect.

MRS. B.

The rays, whether reflected or absorbed by the wall, are alike beneficial to the tree planted against it; for no sooner does the temperature of the wall become elevated by the absorption of heat than it radiates this heat, which is thus transmitted to the tree in contact with it.

The injurious effects of the captivity of the branches must, however, be taken into consideration; for, besides the salubrious exercise which free access of air affords to a plan , this agitation augments the power of evaporation — a power which it is very desirable to encourage, as it is necessarily diminished in a climate of a lower temperature than that in which the plant was placed by Nature;

especially in England, where the atmosphere is impregnated with moisture. It is owing to this circumscribed power of evaporation that wall-trees are less hard in their texture, and contain less carbon, than those which grow freely.

EMILY.

Let me endeavour to recapitulate the several circumstances to be attended to in transplanting from a warm to a cold climate. In the first place, you must make the plant pass through the various gradations of the hothouse and greenhouse, previous to exposing it to the open air; you must then plant it in the spring, deep in a richly-carbonised soil, and cover up the vital point of junction between the stem and the roots. It must be placed in a southern aspect, or against a south wall, watered with great moderation: the fruit must be gathered before the frost sets in; and the plant may be covered with matting in winter, if situated in an elevated spot in a dry soil.

MRS. B.

I believe you have enumerated all the directions I have to give you on this point.

I shall add a few remarks on greenhouses and hothouses, as necessary to the cultivation of such plants as cannot be familiarised to our climate. They should both be situated, as far as is practicable, in a southern aspect.

EMILY.

But when this is not attainable, to which do you give the preference, a south-east or south-west aspect?

MRS. B.

The first has the advantage of affording relief earliest to a plant which has suffered from the cold during the night; the latter that of sheltering it from the severe east wind: upon the whole I should be inclined to choose the latter.

Vertical windows have the advantage of not retaining the snow, the disadvantage of admitting less light and heat; and in England, where we are not much troubled with snow, and require all the heat we can obtain in winter, the inclined windows are certainly preferable, and are almost universally adopted; whilst on the Continent, where less heat is required, vertical windows are more common.

In southern climates, the house must not be built deep, in order to admit the sun's rays to every part; in northern climes, the sun's rays falling more obliquely, a greater depth of building is admissible: the roof should project and be coved, so as to collect the rays of the sun and reflect them into the house.

In order to protect hothouses and greenhouses from humidity, they must be warmed and aired at the same time: the heat both dissolves the moisture and prevents the plants from suffering from the external air while the windows are open; and the current of air carries off any moisture which is not dissolved.

The finest hothouses are in Russia, where the wealth of the higher classes enables them to indulge in luxuries. In their cold climate, hothouses and greenhouses are considered as necessary articles of comfort. In England, Mr. Loddiges' establishment at Hackney exhibits the most perfect model of hot and green houses: they are warmed by vapour, and when necessary, watered by the tepid water into which the steam is condensed. The building is one thousand four hundred feet in length, and divided into compartments, each of a different temperature.

EMILY.

What effect has the tan which is used in hothouses?

MRS. B.

It produces heat, by undergoing a degree of fermentation; but it also generates humidity, and heat and humidity favour the propagation of fungi and of worms; and should the roots of the plant grow below the pot, and penetrate into the tan, its contact is injurious.

The smaller and lower the house, the more favourable it is to young and delicate plants: the heat always rises, so that in large and elevated houses the small plants of the lower range are situated in an atmosphere of cold air. It is for this reason that slips of greenhouse plants are generally placed in beds covered with glass in order to strike root, where they receive as much heat in as small a space as possible.

CONVERSATION IX.

ON THE ACTION OF THE ATMOSPHERE ON VEGETABLES.

MRS. B.

This morning we shall turn our attention to the manner in which the atmosphere affects the vegetable world. It acts on plants in two ways: both mechanically and chemically.

CAROLINE.

But in a very different manner from what it does on animals: we only breathe the air; plants may in some measure be said to feed on it, since they absorb carbon from the atmosphere.

EMILY.

They also absorb oxygen from that source; but it is true they restore it with ample interest, thus purifying the air we animals have contaminated by our breath. But since carbon is such a favourite food of plants, I should like to try the experiment of enclosing a weakly, debilitated plant in an atmosphere of carbonic acid, to see whether the abundance of such nourishment would not restore its vigour.

MRS. B.

The quality of the food would be excellent; but nothing is good when administered in excess. Such an experiment would resemble the attempt made to restore pulmonary patients to health by giving them pure oxygen gas to respire: at first it seemed to be attended with beneficial effects; but the deleterious consequences occasioned by too great excitement of the lungs, was soon discovered, and the experiment abandoned. We cannot be too cautious in our proceedings when we venture to deviate from the paths which Nature has pointed out.

The chemical action of the atmosphere on plants we have already so fully investigated in our preceding Conversations, that, although it be no less applicable to our present subject, it would be but repetition to return to it. I have not, however, yet mentioned, that the electricity of the atmosphere appears to affect plants; it is at least an undoubted fact, that vegetation is accelerated during a storm.

EMILY.

May not that arise from the agitation produced by the wind? Branches being tossed to and fro, must greatly increase the velocity of the sap, and the deposition of its nutritive particles; the evaporation from the leaves must also be considerably augmented by the wind blowing them about, and carrying off the vapour the instant it is formed. Such a state of stimulus might be followed by a debilitating re-action, as is the case with us, after excessive exercise; but, while it lasts, it must produce a very great increase of action throughout the frame.

MRS. B.

That may possibly be a concurring cause of the phenomenon, but is not sufficient to account wholly for it. Mr. De Candolle mentions the remarkable growth of the branch of a vine during a storm, of no less than an inch and a quarter in the course of an hour and a half; now the tree grew against a wall, so as to be little accessible to the wind.

The quantity of water contained in the atmosphere is a point of great importance to plants. Water, you may recollect, exists in the atmosphere in two different states: in the one it is so completely dissolved, that the air feels perfectly dry to us, and affords no moisture to the vegetable part of the creation. It is heat which enables the air to perform this solution; therefore the higher its temperature the more water it can dissolve.

CAROLINE.

Then the atmosphere in the torrid zone, though driest, contains most water: — that appears very paradoxical.

MRS. B.

It is nevertheless true. Whenever the air

cools, its power of retaining water in solution diminishes.

CAROLINE.

This must happen, then, not only when the weather changes from hot to cold, but every evening after sunset.

MRS. B.

Accordingly, we continually see misty vapours floating in the atmosphere in the evening, and the ground more or less covered with dew: all this is water precipitated by the diminution of temperature of the air. In the morning, when the sun has sufficiently warmed the atmosphere to enable it to dissolve these fogs and vapours, they disappear.

The other state in which water exists in the atmosphere is that of a fine subtle vapour, diminishing its clearness, and giving us the sensation of humidity.

EMILY.

The latter state, I should suppose, would be the most advantageous to vegetation, for plants almost always require water.

MRS. B.

It is to the dampness of our climate that we owe the fine grass of our beautiful meadows and lawns, for which England is so celebrated, and which are in vain attempted to be imitated on the Continent; unless it be in some very elevated spots, where the grass is nurtured by the mountain mists. It is to this cause, also, that we are indebted for the prosperity of our laurels, and a variety of evergreens; yet a damp climate has its attendant disadvantages, even as regards the vegetable creation. Fogs and vapour diminish the quantity of light, and, consequently, the numerous benefits resulting from it, such as absorption, evaporation, deposition of carbon, and developement of colour. In this point of view, therefore, a very moist climate injures the beauty and vigour of vegetation.

Trees growing on mountains, where they are much exposed to vapour, are very liable to suffer from what is commonly called a white frost. The clouds and mists, so prevalent in those elevated regions, bedew their branches with a light coating of watery vapour, which easily freezes during the night; the morning mist attaches itself to this thin layer of ice, and shoots into minute frosty crystals, called a white frost.

CAROLINE.

The white frost, which we so commonly see on the grass, is formed, I suppose, by the freezing of the dew.

MRS. B.

Yes; and it is, you know, so slight as to disappear soon after the sun has risen above the horizon.

Moisture is particularly inimical to blossoms: if it comes in contact with the anthers, it destroys them, and the flower bears no seed. This disease often affects the vine, and not unfrequently corn, to the great injury of the vintage and the corn-harvest. Moisture is prejudicial, also, by giving rise to the propagation of fungi, of parasitical plants, and even of worms.

There are some plants to which the moisture of sea-breezes is so essential, that they cannot be cultivated in any other situation than on the shores of the ocean.

EMILY.

These plants, doubtless, require sea-salt; and yet the vapour which exhales from the sea is perfectly sweet, the clouds which they form are never impregnated with salt; how, therefore, can plants obtain it?

MRS. B.

Not from the vapour which rises from the sea, but from minute drops of salt-water, which are projected into the air by the agitation of the waves, and carried by the wind to the shores. Salsola kali, or kelpwort, is a plant of this description: if grown in an inland situation it contains not a particle of soda, the alkali from which it derives its value; for this can be obtained only from *muriate of soda*, or sea-salt. During the late war between France and Spain, the French, being greatly distressed for soda, which they had imported chiefly from Spain, attempted to cultivate kelpwort on the hills bordering the sea-shore in the south of France. When planted on the southern side of the hill, sloping towards the sea, the crop succeeded perfectly; and, the price of soda having risen very high, one single harvest repaid the price of the land on which it was raised. But when planted on the north side, where it was not exposed to the briny particles, it failed completely, and the plant contained no other alkali than potash.

EMILY.

This explanation solves a difficulty which had often perplexed me. The sea-air is, you know, much recommended to invalids as a tonic; and the fogs and damp mists, so prevalent on the sea-shore, do not appear to be attended with the debilitating effects produced by inland fogs: this must be doubtless owing to the tonic qualities of the briny particles with which the sea-air is impregnated.

CAROLINE.

It is true, one can seldom walk out on the sea-shore without one's dress suffering from humidity; and the salt with which this humidity is loaded is often so strong as to be sensible to the taste. But this prevails only within some few hundred yards of the shore: on rising upon the downs it is no longer perceptible; and yet it is the air of these downs which is reckoned so particularly invigorating.

MRS. B.

The saline particles are too ponderous to be carried to such a height; but there you experience the salubrious effect of mountain-air. The salt is not requisite to render it tonic: it is only when you descend on the opposite side into the inland country, that you perceive the want of those invigorating qualities for which the sea-air is so celebrated.

It is remarkable that kelpwort exudes a portion of the alkali which it receives from the atmosphere into the ground, the soil on which it has been cultivated being found to contain more than when such a crop has not been raised upon it. This is owing, probably, to the quantity it absorbs, being more than the plant requires.

EMILY.

This, then, affords an exception to the general nature of the exudation of plants by the roots, as it must constitute appropriate nourishment for other plants of the same species.

CAROLINE.

And is not the air also useful as a vehicle to transport small seeds from one country to another?

MRS. B.

Yes; there is scarcely any resemblance between the plants of Europe and of America, excepting in *Cryptogamous* plants, because the seeds of lichens, of mosses, and of fungi, of which this family is composed, are so small that they float in the air, and are transported by the wind from one continent to the other.

The wind also performs the part of a careful sower, dispersing the seeds which fall from the plant with regularity over the soil.

CAROLINE.

I wish it would distinguish between weeds and flowers, and confine itself to the dispersion of useful seeds; but it seems to delight in the propagation of weeds: if any thistles or groundsel are to be found in a garden, the wind is sure to carry the seeds all over it.

MRS. B.

The distinction between weed and flower is not so easily made as you may imagine. In botany we know not what weeds are; every plant has its use for some purpose or other.

We have already noticed the beneficial effects of the motion which the wind communicates to trees. That of the palm-trees in Egypt is so powerful, when agitated by a high wind, that the French, when in that country, made use of them as levers to draw up water from the Nile. The motion of the stems raised the piston of the pumps, which fell again by their own weight.

Of the temperature of the atmosphere, which is the point of greatest importance in vegetation, we have already fully treated. I have only to add, that plants which grow in the plain, in countries of high latitudes, if transported to a warmer climate, must be cultivated in elevated situations. In Chili, for instance, potatoes grow at an elevation of nine thousand feet higher than they will grow in these climates.

EMILY.

The degree of latitude is, then, the inverse of the degree of elevation from the level of the sea?

MRS. B.

The one serves as a compensation for the other. It is not, however, every species of plant which can take advantage of this sort of compensation; but the greater number of plants will grow equally well at a high latitude in the plain, or in a low one on a mountain.

In an estimate made of the greatest height at which several different species of trees will grow in the south of France, it appears that

The Larch grows at an elevation of 7200 feet.
The Birch .... 6000 feet.
The Beech.............................. 4000 feet.
The Cherry .... 3000 feet.
The Walnut .... 2400 feet.

Among the *Cerealia*,

Rye .... 6000 feet.
Wheat.............................. 5400 feet.
Turkey Corn.............................. 3000 feet.
The Vine 1800, or even so high as 2400 feet, if the situation be particularly favourable.

EMILY.

I recollect woods of birch-trees on the mountains of Scotland, where they had to struggle against the difficulties both of elevation, of latitude, and of situation; but, it is true, they were ragged and dwarfish, and wore the appearance of great distress.

MRS. B.

And yet their elevation was far below 6000 feet; for Bennevis, the highest mountain in Scotland, does not rise more than 4370 feet above the level of the sea. The larch, which is much more hardy, has been planted upon the mountains of Scotland with great success, and clothes their once-barren sides with a delicate foliage.

The olive-tree will not grow in a higher latitude than the southern provinces of France; and there it is only under the most favourable circumstances of soil and aspect that it can be cultivated, at the height of 1200 feet.

CAROLINE.

We have seen it in Italy growing almost to the summits of the mountains. I own that I was disappointed with the Italian olive. We associate so many pleasing and poetic ideas with the olive-branch, that I expected it to partake of the beauty of its rival in the gardens of Parnassus — the laurel; but, so far from vying with it in beauty, the colour of the olive is so dingy, that it looks like a willow covered with dust.

MRS. B.

I cannot see what reason there is to expect much similitude of appearance between the emblems of glory and of peace. The branching of the young olive is, however, remarkably elegant, and the lightness of the foliage atones for the want of vivacity in its colour; and, when mixed with plants of a more lively green, affords a very agreeable variety. The olive-groves of Tivoli gave me the impression of the most antiquated trees I had ever beheld: their venerable trunks are torn, riven, and twisted into a thousand fantastic forms; a profusion of young branches, decked with light foliage, shoot from these aged stems, and, waving their silvery tints in the sun, seem to smile and say, We are your contemporaries; but the antiquated parent whence we spring once put forth branches under the shade of which Cicero and Mecænas reposed.

CAROLINE.

They are the only olive-trees I ever admired. Independently of their magnitude, they are interesting from their appearance of age and decrepitude. The stems, so curiously split asunder, look as if they bid defiance to the violence they have suffered; and, twisting the dismembered remnants of their stems together, seem to verify Æsop's fable of the bundle of sticks, and acquire additional vigour to withstand the attacks of time and of the elements.

CONVERSATION X.

ON THE ACTION OF WATER ON PLANTS.

MRS. B.

Water may be considered as acting on plants in several different ways: the first and most important of which is, its being the vehicle of their nourishment. The greater the quantity of nutritive particles which water contains the more favourable it is to vegetation, unless it should be so far saturated as to be too dense to pass through the pores of the spongioles. In that case the plant is reduced to the state of Tantalus, and perishes of famine in the midst of plenty.

We have already observed, that, of three particles of water which enter a plant, one only remains within it; and this either retains its natural liquid state, analogous to the water of crystallisation in minerals, or is decomposed, to contribute to the formation of oils or other peculiar juices of vegetables.

In the second place, water acts mechanically on plants, by dilating them, and rendering them supple. The woody fibre absorbs water abundantly, but not the bark. The former is often so swelled by this absorption as to burst and split the bark.

CAROLINE.

This is no doubt one of the causes of the roughness of the bark of many trees, such as the oak and the elm, which are so seamed and severed into small parts.

MRS. B.

Yes; these, in the course of time, dry and fall off. In other trees the bark remains smooth, but peels off when split by the swelling of the wood.

EMILY.

But how can the wood absorb water if it does not pass through the bark?

MRS. B.

The wood absorbs water by the internal vessels. If the trunk of a felled tree lie on the ground in a damp spot, with the roots and branches cut away, so as to leave the vessels exposed at both ends, it will absorb so much water as frequently to make it sprout small branches and leaves.

EMILY.

I recollect seeing an instance of this in Kensington Gardens: the log was lying in a dell, partly immersed in water, and the whole of it was sprouting with verdure.

MRS. B.

Thirdly, water conveys air into plants; and the more it is impregnated with air, whether atmospheric or carbonic acid, the better it is adapted to vegetation. Thus the water of large rivers which flow rapidly, and pass over a great extent of country exposed to the influence of the atmosphere, is much more favourable to vegetation than that of small rivulets, which have not been a sufficient length of time in contact with the atmosphere to become impregnated with air: yet this latter is preferable to the water of a lake, which has no current; for it requires motion, and pretty strong motion, to mix water and air together in the quantity which is required by plants. Large lakes, such as that of Geneva, it is true, are considerably agitated by the wind: the waves then swallow up a certain quantity of air; but when tranquil it contains much less than river water.

CAROLINE.

Yet is not the water of ponds preferred to that of rivers for watering plants, although it is so tranquil as frequently to become stagnant, and to be covered with a green slime, which shows that it can have been little agitated by the wind?

MRS. B.

This green scum is a vegetable production, affording food to numerous swarms of the insect tribe, flies, worms, snails, &c. In a short space of time this little ephemeral population, as well as the vegetables which nourished it, perish, and putrefaction succeeds, as you may frequently have discovered by the offensive effluvia exhaled by ponds and marshes of this description; but the corrupted waters, disgusting and deleterious as they are to us, afford a feast of abundance to the vegetable creation. Saturated with the decayed remnants of both animal and vegetable matter, and replete with carbonic acid, they convey these rich materials for fresh vegetation into the roots of the living plant.

CAROLINE.

The only danger, then, is lest the fluid be too abundantly laden with food, and the plant be gorged with it.

MRS. B.

As plants are not capable of acquiring the virtue of temperance, Nature has wisely provided against their suffering from excess, by giving them mouths of such very small dimensions, that they cannot take in more than is good for them. The only danger, therefore, is, lest the fluid should be too dense to obtain entrance at the roots.

The water least appropriate to plants is that of springs; and, when obliged to use it, we should endeavour to remedy the double defect of a deficiency both of air and of temperature, by leaving it exposed during some length of time to the atmosphere. If the reservoir in which it is contained be situated in the neighbourhood of a farm-yard or stables, the water will become impregnated with carbonic acid evolved by the manure. Advantage may also be taken of such a vicinity to enrich the water of the reservoir, by conveying a small stream through the manure into it.

EMILY.

But spring-water is not always of a lower temperature than the atmosphere: during a frost it is evidently warmer, being sheltered from the cold by the depth whence it rises.

MRS. B.

Plants do not then require water: it is in summer that this artificial aid is wanted, and never at a season when the temperature of spring-water is higher than that of the atmosphere.

EMILY.

One cannot sufficiently admire the beautiful provision which Nature has made for watering her vegetable creation. The rain, falling in small drops through the atmosphere, acquires its temperature, and becomes impregnated with air.

CAROLINE.

Yet rain-water can contain no nourishment, unless it be a little carbonic acid it may imbibe from the atmosphere in passing through it; for rain, consisting of pure vapour exhaled from the surface of the earth, can hold no nutritive particles in solution.

MRS. B.

Observe that, when rain falls on a plant, it merely refreshes the foliage, by washing off the dust, and cleansing the evaporating pores, which may have been clogged during the drought. Rain cannot feed the plant as it falls from the clouds; the absorbent pores, you know, are not exposed to its influence: in order to perform this second function it must penetrate into the earth in search of food, and, dissolving whatever it meets with appropriate for that purpose, convey it to the roots of the plant.

Attention must be paid to the proper time and season for watering plants. They do not require it at all in winter, when growing in the open air; because, during that season, they cease growing, and, consequently, stand in no need of nourishment; indeed, they often absorb more water from the wet soil in winter than is good for them. Greenhouse and hothouse plants should be watered with great moderation in winter: it is time enough to supply them when they ask for it, which you may perceive by their leaves beginning to droop and wither. As spring approaches, the quantity of water must be increased, in order to feed the young buds, which call up additional sap; but the increase must be made with precaution, the earth being still moist with the winter rains. Plants at this season should be watered in the morning, in order that they may not be overloaded with moisture during the night, which would be dangerous should a frost chance to occur; besides, by watering in the morning, you provide for the evaporation of the day.

EMILY.

Yet our gardener generally prefers watering in the evening, if he does it only once a day.

MRS. B.

In the midst of summer the plant, exhausted by evaporation during the heat of the day, requires water in the evening to revive it; there is then no danger of its suffering from frost during the night. In Switzerland, where heat and light are much more powerful than they are with us in England, it is generally necessary to water plants both morning and evening: the earth is dry; and it is difficult in summer to provide for the immense increase of the absorbing and evaporating functions.

There are some plants which grow perfectly well on the Alps, because they are, throughout the summer, watered by streamlets supplied by the melting snow: these same plants perish on the Jura, or any other mountain which is free from snow in summer, because they are not furnished with so regular a current of water.

In autumn, trees which bear pulpy fruits, such as the peach and the plum, require a great deal of water to fill out and to ripen the fruit. Fruits of a dry nature, such as nuts and dates, do not need so much; and the precaution of watering in the morning is equally necessary as in the spring, lest the plant should be surprised by a frost during the night.

CAROLINE.

Grasses and herbs, I suppose, require more water than trees; for, consisting chiefly of leaves, they must undergo a greater evaporation.

MRS. B.

Yes; and annual grasses the most of any.

Seeds which are beginning to germinate, should be watered very sparingly; for the seed, feeding at first on its own proper substance, is rather in want of air than of water; but as soon as it has put forth roots, and a stem has sprung up, it will require a more plentiful supply until the time of flowering, when it must again be restricted, because the blossom is nourished by its own peculiar juices elaborated by the leaves; and when the seed ripens, if it be of a dry nature, still less water must be given.

The quantity of water depends, also, upon the nature of the cultivation. You must consider what is the produce you wish to favour, and water accordingly: if it be a meadow, leaves and not flowers will be your object; therefore you must water profusely, since abundance of water is favourable to leaves, and prejudicial to flowers. If it be a field of corn, it is the grain you would favour; therefore you must water sparingly. Rye is cultivated sometimes with a view to the grain, and sometimes chiefly for the straw; in the first case you must water but little, in the latter abundantly.

The siliceous soil of Ireland is very favourable to the culture of corn: this earth not being retentive of water, the abundant rains of that country do not injure the crops. A similar soil in France will not admit of the cultivation of corn; because the climate being much hotter, the corn requires more water, and can be raised only on an argillaceous or clay soil, which retains the water.

EMILY.

In watering fruit trees, I have observed the gardener dig a trench round the tree, at some little distance from the stem, and pour the water into it, instead of watering close to the tree.

MRS. B.

A judicious gardener will apply the nourishment to the mouths of the plant it is to feed. Now these, you know, are situated at the extremity of the roots; and, as the roots spread out beneath the soil, pretty nearly to the same extent as the branches above ground, the tree should be watered at the distance of the extremity of the branches from the stem; the closeness of garden culture usually prevents a trench being dug so far from the tree, but the nearer you approximate to it the better. Observe how admirably Nature teaches us this lesson: the head of the tree, in the form of a dome, protects the stem from the rain, like an umbrella: all around the soil is exposed to the rain, and the water penetrates the earth just where the extremities of the roots are situated to receive it. In addition to this, the greater part of the rain, which has washed and refreshed the leaves, trickles down from the ends of the branches, and reaches the ground in the appropriate spot.

CAROLINE.

How beautifully contrived! I shall not in future take shelter from a shower, beneath a tree, without thinking of it.

EMILY.

What strikes me with the greatest wonder, in these arrangements of Nature, is the ease and simplicity of the means employed: it is always a natural consequence — a thing of course; it would require efforts to prevent, rather than to produce such results: the facility with which they are accomplished, does not draw our attention; but when we do observe and study them, we cannot but feel their infinite superiority to the most complicated contrivances of art.

MRS. B.

The greater and more comprehensive the mind that contrives, the more simple, in general, are the means employed; you may admire, therefore, but you can scarcely wonder at the perfection of the economy of Nature.

In order to form correct ideas on the theory of watering, we must distinguish between the means which are natural and those which are artificial. The former consists in rain, dew, and the melting of snow. Since it is beyond the power of science to augment or diminish the quantity of rain by a single drop, or to accelerate or retard, by a single minute, the period of its falling, we must, with great humility, limit our efforts to the study of the signs of the times and seasons of approaching rain, in order to modify our culture, so that it shall receive advantage and not injury from it.

EMILY.

Does not the barometer indicate the approach of rain with tolerable accuracy?

MRS. B.

Far from it: according to the most exact calculations, it is found that the descent of the mercury is followed by rain only seven times out of eleven.

CAROLINE.

Then you have the hygrometer?

MRS. B.

That is of little use as a sign of approaching rain: it indicates merely the degree of moisture of the spot in which it is situated, and gives us no in sight into the state of the upper regions of the atmosphere.

Wind blowing from places where a greater degree of evaporation takes place is one of the most unquestionable precursors of rain. This is the case with winds blowing from the sea; thus the west wind comes loaded with vapour from the Atlantic Ocean, which it deposits on the continent of Europe.

EMILY.

But why does the south wind bring us rain; we may consider that as coming from the dry heated continent of Africa, for the Mediterranean Sea is too insignificant to impregnate it with vapour?

MRS. B.

The climate of Africa being considerably hotter than that of Europe, a greater evaporation takes place there; the atmosphere dissolves and contains much more water than our colder regions are capable of holding in solution; the air, therefore, as it advances northward, becomes loaded with a precipitation of vapour, which congregates into clouds, and falls to the earth in the form of rain.

CAROLINE.

That is very curious; and a north wind, on the contrary, being able to maintain more vapour in solution in our climate, than it did in the colder countries whence it blows, scarcely ever brings us rain. I have heard that swallows flying low, flies stinging, fowls rolling themselves in the dust, and cattle feeding voraciously, are all signs of approaching rain; pray, are these merely rustic prejudices, or will they admit of an explanation?

MRS. B.

Swallows fly low before rain to catch the insects, which then come nearer to the earth for shelter; they may, also, approach the earth in search of worms, which make their appearance above ground in times of rain; then a species of fly, with an indurated trunk capable of inflicting a wound, frequently makes its appearance on the approach of wet weather; fowls may possibly cover themselves with dust in order to preserve them from the wet; and cattle may, instinctively, lay in a store of food as a provision against the time they must abandon their pasture to seek shelter from the rain; but I do not pretend to advance these opinions as any thing more than conjecture founded on some appearance of plausibility.

The second mode which nature employs to water plants is the dew. I hope you recollect the very ingenious theory of Dr. Wells on that subject.

EMILY.

I fear but imperfectly.

MRS. B.

I advise you to look it over *; at present I shall only say that it is founded on Professor Prevost's Theory of Radiant Heat. In proportion as a

* See Conversations on Chemistry, vol. i. p. 99.

body radiates, its temperature must necessarily be lowered, unless it be supplied with heat from some foreign source: during the day the sun affords this supply very amply, but after sunset the earth, as well as every object upon its surface cools by radiation. The atmosphere, which radiates much less than the solid earth, preserves its temperature longer; but the stratum of air which is immediately in contact with the ground is cooled by it, and deposits upon it that portion of vapour which the diminution of its temperature prevents it from longer holding in solution. This precipitation is the dew, which you perceive on the grass, after sunset.

EMILY.

Since it proceeds from the cooling of the surface of the earth, why is it not equally precipitated on gravel walks and pavement?

MRS. B.

Because the stones of which these are composed are not good radiators, and therefore preserve their temperature longer; and if they do not cool quicker than the air with which they are in contact, no deposition of dew will take place. Minerals, and especially metals, are bad radiators; they require no dew: Nature reserves this mode of watering for the vegetable creation: to plants she gives the power of abundant radiation, both to enable them to throw off the heat with which they have been oppressed during the day, and to call down those refreshing showers of dew which restore their vigour. One knows not which most to admire, the wise provision which is thus made for the benefit of the vegetable kingdom, or the simplicity of the means by which it is accomplished.

CAROLINE.

I imagined that the dew fell from a considerable height; for trees afford a shelter from it: you seldom find any dew beneath a tree?

MRS. B.

The radiation of the earth is stopped by the canopy of the tree and reflected back to the ground, thus preventing it from so rapidly cooling as to occasion a deposition of dew. For the same reason, when the sky is covered with clouds, the heat is reflected back to the earth by them, and little or no precipitation of dew takes place; while, on a clear night, the radiation goes on uninterruptedly, the earth cools rapidly, and an abundant dew is deposited.

EMILY.

How admirably this provision is proportioned to the wants of the vegetable creation! A clear sky, which leaves it exposed throughout the day to the ardour of the sun's rays, insures it an abundant supply of refreshing dew in the evening.

CAROLINE.

I have seldom perceived this radiation of heat

in England; but in Switzerland it is very sensibly felt on a summer's evening, from trees, walls, and other buildings which have been heated by the sun during the day.

MRS. B.

It is for this reason, that, in hot climates, the public walks are less planted with trees, than those of more temperate regions; in the former you can walk out only after sunset, when the neighbourhood of trees is attended with every disadvantage. They prevent the free circulation of the cool evening air. They reflect back the heated radiation of the earth, and are, themselves, a source of heat by their own radiation.

EMILY.

In our more temperate climate, when we frequently walk out during the day, trees afford us a grateful shelter from the sun, and in the evening they have the advantage of retaining the heat and preventing the deposition of dew.

MRS. B.

In regard to the third mode which nature employs to water plants, the melting of snow, as it relates only to plants growing on the Alps, or other mountains whose summits are constantly covered with snow, it is unnecessary to make any observations upon the subject.

CONVERSATION XI.

ON THE ARTIFICIAL MODES OF WATERING PLANTS.

MRS. B.

We shall now proceed to examine the artificial modes of watering, which may be divided into three classes.

1st. By watering-pots or engines.

2d. By filtration.

3d. By irrigation.

The first mode applies merely to horticulture, for the use of watering-pots can scarcely be extended beyond the garden and greenhouse; the plain spout is calculated for watering the roots, that pierced with holes for washing the leaves, and for watering seeds, young sprouts, or delicate plants which require to be watered sparingly.

If the soil be light, or the plants situated near the high road, or exposed to the smoke of a town, they require more water to refresh the leaves; for if the stomas are choked, evaporation is checked, and vegetation injured: in order to render your plants healthy, they must not only be well fed, but kept clean.

EMILY.

Green-house and hot-house plants being sheltered from the dust, will, I suppose, not require so much precaution?

MRS. B.

On the contrary, they are exposed to the dust which arises from their cultivation within doors, and deprived of the natural means of getting rid of it, the wind, which in the open air prevents the dust from accumulating on plants : this artificial mode of raising plants, therefore, requires more attention to cleanliness than when grown in the open air; and gardeners frequently use a bellows as a substitute for the wind.

CAROLINE.

Is it not a good way of keeping greenhouse plants moist in the summer to bury them in their pots in the earth?

MRS. B.

Yes, provided they are taken up occasionally, in order to cut off the roots, which shoot through the aperture at the bottom of the pot; for if this operation be delayed till they are housed in the autumn, the roots will be so bulky as to render their amputation dangerous.

CAROLINE.

Yet is it not the nourishment which the plant

obtains from the soil, by shooting its roots through this aperture, which gives it so much vigour?

MRS. B.

True, but the small fibres which sprout out after cutting away the projecting roots are sufficient for this purpose. The main object of the aperture at the bottom of the vase, is in order that the water may filter through; without this resource it would become stagnant around the roots, and rot them The opening is, you know, partially closed by a piece of tile, leaving not more room than for the water to escape which is not absorbed by the roots; but when water is not supplied to the plant in sufficient quantity, the fibrous roots insinuate themselves through the aperture to search for it in the soil beneath.

Watering by filtration is adapted to two classes of plants; those which suffer from excess, and those which suffer from scarcity of water: it may be performed in two ways, the one is by enclosing the vase which contains the plant, in a larger one full of water, and then burying the double case in the earth, the water will filter from the outer into the inner vase, which must not, of course, be glazed, but of a porous texture. The other mode is to place a pot of water contiguous to that which contains the plant, and connect them by means of a skein of worsted which will act as a syphon, and transfer the water to the vase in which the plant grows.

In the Isle of Corfu I have heard that it is usual to water the orange trees by surrounding them, at the distance of the extremities of the roots, with very porous pots of water; the water oozes through into the ground, and is sucked up by the spongioles.

Meadows are commonly watered by filtration, small trenches are dug, into which water is occasionally made to flow, and thence it filters into the adjacent soil; these trenches should be rather below the surface of the soil, in order that the water may the more easily penetrate to the roots of the grass.

EMILY.

The trenches are, I suppose, left open in order that the water may derive the benefit of exposure to the air?

MRS. B.

They are sometimes buried in the soil, and at others left open. The first have the advantage of economising the soil, as the ground above them may be cultivated; loss is also prevented by evaporation; yet I prefer the open trenches, both on account of exposure to the air and as affording facility for repairs, which are often required.

3dly. Watering by irrigation consists in conveying the fluid through small channels similar to those used for watering by filtration, but which are made, at pleasure, to overflow the adjacent ground. In order to accomplish this, it is necessary to be furnished with an ample supply of water; if it can be obtained from a superior elevation the operation is greatly facilitated.

When, on the contrary, it is necessary to raise it from rivers or wells, various mechanical means may be resorted to. The current of a river may be used to turn a wheel furnished with small buckets, which, during one revolution of the wheel, fill with water, raise it, and pour it into the reservoir prepared to supply the rivulets of irrigation: when there is no current, horses may be employed to turn the wheel.

The hydraulic ram is another mode of raising water. M. De Candolle mentioned one, which, put in motion by a fall of water of twenty feet, raises a body of eight cubic feet per minute to the height of one hundred and sixty feet.

EMILY.

I should think a steam-engine would afford the most effectual means of raising water; is it not used for this purpose?

MRS. B.

Very frequently, for draining mines; but it would, I conceive, be too expensive a mode of raising water for the purpose of agriculture; at least, I never heard of it being so applied.

It is to be regretted that the rain water which is washed down from the roofs of houses should not be turned to account; soiled as it is by this operation, it would but be the better calculated for the nourishment of plants; and it might easily be collected into a reservoir, instead of being carried off, as it usually is, by the common sewer.

It is singular that we should have first learnt artificial modes of watering from the Moors of Spain; their labours in that department were very extensive. Near Alicant they constructed a wall between two hills in order to retain the water which flowed through the valley, for the purpose of irrigating the adjacent country. This wall, which is still in existence, is only twenty-four feet in length at the base, this being the breadth of the valley; but the hills receding as they rise, it is two hundred and sixty feet long at the top, and sixty-seven feet in depth; which is much more than is required to withstand the force of the waters it confines; but the Moors were not versed in the laws of hydraulics.

There are various modes of irrigation: the inundations are sometimes flowing, sometimes stagnant; sometimes transitory, at other times permanent, according to the nature of the culture. Of the latter description are the rice plantations: this plant requires such abundance of water, that the inundation is drawn off only when the grain begins to ripen.

EMILY.

I remember, in Lombardy, seeing the green tops of the rice peeping through their watery bath, and looking not very unlike the green scum which frequently covers pools of stagnant water. Nor does it appear to be less pernicious; for I have heard that the cultivators of these rice fields are often afflicted with a frightful cutaneous disease, which terminates frequently in madness and self-destruction.

MRS. B.

The cultivation of rice is certainly not a healthy employment, owing to the stagnant waters in which it is raised; but the disease to which you allude, called the *Pelagra*, is supposed to proceed from feeding on maize, or Indian corn, improperly prepared. The origin of this dreadful malady was for a long time an inexplicable mystery; and it is only lately that Dr. Sette having observed that it was confined to those districts in which the maize, instead of being preserved in the ear, was kept, like wheat, in separate grains, ready to be ground into flour. This led him to suspect that the grain might have acquired some deleterious property; and, on examining it with a microscope, he discovered that that part of the grain, by which it had been attached to the husk, was covered by a species of mould of a poisonous nature; there is, therefore, every reason to believe that this fatal disease arises from feeding on maize in this corrupt state; and if so, the disease might be easily guarded against.

EMILY.

What a fortunate discovery! The remedy is so simple, it is merely to adopt the usual mode of preserving maize in France and Switzerland, by hanging it up to dry in the ear.

MRS. B.

Certainly. We have hitherto considered only the various modes of administering water; but it sometimes happens that the earth is too moist: it is necessary, therefore, for the purposes of agriculture, to be acquainted with the best mode of draining it. This operation may be performed in several ways. When the locality will admit of constructing a reservoir in a lower situation to receive it, the water may be carried off by subterraneous ducts. These conduits should be filled with pebbles, sufficiently large to leave a free passage for the waters between them.

CAROLINE.

But of what use are the stones? why not leave the channel quite free and open?

MRS. B.

The stones may be considered as forming a sort of loose wall which serves to support the duct, by preventing the top and sides from falling in: the water would soon wear them away, were they not thus defended, and the passage be obstructed.

The operation of draining a marsh is of much greater importance. A marsh is a space of ground on which the water remains too long, either for want of means of running off laterally, or because a layer of clay soil prevents it from filtering downwards.

CAROLINE.

I should have thought that this abundance of water would have been favourable to the culture of many species of plants.

MRS. B.

It is true that marshes cannot be said to be inimical to vegetation in general; for these spots abound with plants, but they are of an aqueous nature, which are good neither for man nor cattle. They afford, however, an ample repast for the creeping things of the earth; and when we condemn noxious weeds and stagnant marshes, we should remember that, though man is lord of the creation, this world was not made for him alone, and that the reptile and the worm have also their share of its enjoyments. When it is required to drain a marsh of small extent, it may be done by planting it with willows, alders, and poplars. These trees being of very rapid growth absorb a considerable quantity of water, the greater part of which they evaporate into the atmosphere. The poplar has also the advantage of affording very little shade; it does not, therefore, interfere with the action of the sun and air, agents which perform very prominent parts in the operation of draining.

Very extensive marshes will not admit of being drained merely by planting. In this case, the mode resorted to is to raise a bank of earth around the marsh, which answers the purpose of a dam, and prevents the water from running into it; for, if once you accomplish this, the marsh is soon dried by the mere process of evaporation. If the marsh be occasioned by a clay soil, the argillaceous earth, which does so much harm by retaining the water after it has entered the marsh, will do as much good by its impermeability when raised in the form of a bank to prevent the water from entering. In digging for this purpose, the earth must be thrown up towards the marsh, so as to leave the trench or ditch external; the water will then run off by this ditch instead of filtering through the dam raised on the other side. The dam should be formed in the shape of a hog's back, and pressed down hard towards the base, in order to prevent the water in the ditch oozing through. The dam or dyke may be planted with trees, the roots of which will help to keep it together, and the evaporation by the leaves will assist in draining it; but care must be taken to thin the branches, in order to give free access to the sun and wind; nor must they be allowed to grow high, lest the wind, having too great hold of them, should loosen the roots, and thus injure, instead of preserve, the dyke. On the side next the ditch it is advisable to plant reeds.

EMILY.

But how are you to get rid of the water which fills the ditches?

MRS. B.

That depends, in a great measure, upon the locality: it must be carried off to the nearest running water, or to the sea; without a resource of this kind, it would be vain to attempt to drain a marsh. If we can succeed in preventing the external waters from gaining admittance, that which the marsh contains is so soon dried up by evaporation, that care must be taken not to overshoot the mark, and leave an insufficiency of moisture for the purpose of cultivation.

CAROLINE.

Some attempts of this kind appear to have been made towards draining the Pontine marshes; a canal of water borders each side of the road, which is flanked by a bank of earth planted with trees; and when we passed, I saw, with regret, that they were cutting most of them down.

MRS. B.

It was probably found, that more injury was produced by their shade, than benefit derived from the evaporation of their leaves. But this attempt at draining the Pontine marshes is of a very circumscribed nature, and attended with little success; although the vicinity of the sea affords facility for carrying off the water, the difficulty of draining these marshes has never yet been surmounted.

When marshes are situated below the level of the sea, as is generally the case in Holland, to drain them is a very laborious undertaking, and requires all the patient persevering industry of the Dutch to accomplish. They begin by making the water run off into canals, and then raise it, by mechanical means, into more elevated channels, till it attains an elevation above the level of the sea.

CAROLINE.

But the level of the sea varies according to the tides, being many feet more elevated at high water than at ebb tide?

MRS. B.

The medium must therefore be taken as a general level; and, raising the most elevated canals above that, let off the water at ebb tide into the sea: the canals are furnished with locks, which are then closed, to prevent the water returning when the tide flows. The mode used in Holland to raise the water from the lower to the upper canals consists of a species of small windmills: as the wind blows regularly, though not violently, in that country, they perform their office very well. The same process of windmills is adopted in Cambridgeshire and Bedfordshire.

EMILY.

Pray can you explain to us the mode in which the valley of Chiàna in Tuscany, which was anciently an unwholesome marsh, has been brought to such a beautiful state of cultivation as it now exhibits?

MRS. B.

The marshes in Tuscany are formed by the

waters which flow from the Apennines, and which, not finding a sufficient vent, are arrested in their course, and become stagnant. The Apennines being of a loose sandy texture, the waters bring with them a great quantity of earth, which they deposit in the lowest parts to which they flow.

CAROLINE.

I recollect that the Arno, when swollen by rain, is quite thick with mud, brought down from the mountains by the torrents which feed it.

MRS. B.

The celebrated Torrecelli took advantage of the deposition of this mud to invent a mode of draining the marsh of the Chiana. But before I explain the remedy, it will be necessary to inform you whence the evil arose, and give you an account of the origin of this marsh.

In ancient times, the numerous rivulets which flow from the adjacent hills into the valley of Chiana poured their united waters from thence into the Tiber. Some inconvenience being experienced by the Romans, from occasional overflowing, they constructed a dyke to close this outlet. The waters, thus stopped in their course, formed a lake, which, when raised by accumulation to a certain elevation, found an issue to the north, precisely in the opposite direction to that in which the water formerly flowed out of the valley. The rivulets, therefore, on entering the valley, were all obliged to change their course. In making this turn, their velocity was diminished; and having no longer power sufficient to carry with them the earthy materials with which they were laden, these were deposited in the lake, in which they accumulated, and, in the course of time, converted it into a marsh. The ingenious and sagacious Torrecelli availed himself of the evil to devise a remedy; and employed the very means which had converted the lake into a marsh to convert the marsh into dry land.

EMILY.

That was a most happy idea; but how did he accomplish it?

MRS. B.

He caused a mound or bank of earth to be raised towards the base of the hill, around the part where a rivulet changed its course. This was left open on the most elevated side; so that the water, laden with earth, in its descent might have free access to it; on the lower side a small aperture was made through which the water alone could escape, leaving behind the earthy matter, with which it was saturated. In the course of time, the soil within this enclosure, was elevated by the accumulation of earth, above the level of the stagnant waters; and rose, like a dry little island, on the edge of the marsh. A contiguous enclosure was then made, and raised by similar means; a third and a fourth followed in succession. These labours have been going on during two centuries: of late years they have been prosecuted with great activity and sagacity by the celebrated Fosombroni of Florence; and only a few years more will be required to complete them. Already you have beheld this district transformed, from a melancholy and pestilential marsh, into a richly cultivated valley, watered by a clear stream, the result of the torrents purified from their earthy deposits.

CAROLINE.

It is, indeed, quite a metamorphosis; and is not this mode adopted in other countries?

MRS. B.

It is frequently employed in the environs of Bologna, and in several other parts of Italy. This operation is called in Italian *colmare*, in French *combler*, that is to say, to *fill up*. I once, in travelling, saw it carrying on, in a spot on the declivity of a hill; for you understand that it can take place only where the ground slopes, so as to enable the waters to run off.

CAROLINE.

When the mountains, from which the rivulets bring down the earth are of schist, like the Apennines, this operation must be much more easily effected than when they are of granite, for the harder the earth the less earthy matter the waters can wash down.

MRS. B.

When the mountains are of granite, no deposition of earth takes place to interrupt the course of the streams, and produce a marsh: the evil cannot exist, and the remedy is not required.

EMILY.

But elaborate and artificial as this mode appears, it is, in fact, precisely that which Nature employs to level the inequalities of the globe: the streams are ever conveying earth from the mountains to deposit it in the vallies, thus lowering the one and elevating the other.

MRS. B.

That is perfectly true: it is thus that the plains in the north of Italy, between the two ridges of the Alps and the Apennines, have been formed. The rivers flowing from these long chains of mountains have deposited their solid contents in the intervening low lands, raised and united the several vallies, and levelled them into plains, such as those of Lombardy and Liguria; and, had Nature been allowed time to complete her work, they would have been elevated to a height which would have preserved them from danger; but impatient man was eager to inhabit this alluring paradise, before its creation was completed. Hence, instead of profiting by the gratuitous labours of Nature, who was gradually preparing it for his reception, he has been compelled to repair by artificial means, at the expense of immense toil and trouble, the evils resulting from the interruption given to her operations.

EMILY.

But of what nature are those evils?

MRS. B.

Inundations produced by the quantity of turbid waters, which in rainy seasons, is frequently so great, as to overflow the whole country, and destroy cultivation. The inhabitants, therefore, found it expedient to put a stop to this levelling system of Nature by embanking the rivers, in order to confine the waters within their beds.

CAROLINE.

We observed that in Lombardy and in Tuscany the rivers were generally embanked: but I should have thought that such a measure would have afforded but a temporary remedy; for those very sands, which Nature would have employed to raise the general level of the plains, being deposited at the bottom of the rivers, would in the course of time, so raise their beds, that the waters would overflow the embankments.

MRS. B.

Very true; their only resource was to raise the embankments in proportion as the bed of the rivers were elevated. In consequence of this elevating system of art, in opposition to the levelling one of Nature, the Adige and the Po are higher than the plains which separate the two rivers; and it is thought that it will be ultimately necessary to form new beds for their waters, in order to avoid the ruin they threaten.

The plains of Holland derive their origin from a similar process; but they are exposed to still greater dangers than those of Italy, lying so low as to be menaced not only by the overflowing of the Rhine and the Moselle, from the shallowness of their beds, but by inundation of the sea. Every defence which art can afford, such as embankments, dykes, canals, &c., has been achieved by the patient and industrious inhabitants of that enterprising country; yet the resistless ocean frequently breaks in upon them, and destroys all their labours.

EMILY.

What a prodigious quantity of earth and sand these rivers must carry into the sea! It is well that its bed is too deep, to be affected by such depositions.

MRS. B.

It is true there is no danger of their occasioning an overflowing of the sea: important effects, however, are frequently produced on its shores. The impulse of the rivers is diminished on reaching the sea, by that of the waves they have to encounter. Sometimes their waters are partially repelled back on the shore, where they form marshes: districts of this description abound on the coast of the Adriatic. Sometimes they deposit their solid contents in one large bank, when their current is first repulsed by the waves of the sea, and form at the mouth of the river a plain or Delta: such is the Delta at the mouth of the Nile.

CAROLINE.

And if we may be allowed to compare small things to great ones, such I suppose is the origin of the Delta or low land, at the mouth of the Rhône, on entering the Lake of Geneva from the Valais. And the Isle of Camarque, formed by the deposition of the waters of the Rhône at its entrance into the sea, offers a still more striking example.

MRS. B.

Sometimes, the rivers have sufficient power to struggle against the resistance of the waves, and do not deposit their mud and sand, till they have advanced to some little distance in the sea; when their waters, broken and divided by the waves, precipitate their cargo in separate spots, forming a number of small islands; hence the origin of those on which Venice is built.

We have prolonged our conversation rather beyond bounds to-day; I have but one more remark to make to conclude this subject.

When it is required to resist the force of an irregular mountain-torrent, a number of small embankments is preferable to one large dyke, however strong; for during violent rains there is some danger of the dyke being carried away, while several small embankments successively break the force of the waters.

CONVERSATION XII.

ON THE ACTION OF THE SOIL ON PLANTS.

MRS. B.

Our last conversation was upon water: to-day we shall change the subject to dry land. It has been asserted, that earth was not absolutely essential to vegetation, because there are some plants which do not require it, which live in water, whence they derive their nourishment: but this class is very insignificant. The earth affords both support and nourishment to plants.

CAROLINE.

Or should you not rather say, is the vehicle of their nourishment, since their food is composed principally of animal and vegetable remains?

MRS. B.

Very true; the various saline particles which plants pump up from the soil, should rather be considered as flavouring their food, than forming a nutritive part of it: their daily bread is of animal and vegetable origin.

CAROLINE.

And when there is a deficiency of salts to flavour their food, have not plants the power of forming them in their internal laboratory?

MRS. B.

No; the chemical apparatus of their organs is so arranged that it can elaborate only vegetable juices, and is as incapable of forming a salt or an oxide, as an animal is of forming the phosphat of lime, with which its bones are indurated.

EMILY.

How then are these salts, which are composed of various ingredients, formed?

MRS. B.

The metals intermixed with the earths of which our globe is composed attract oxygen from the atmosphere, and combine with it; and it is thus that the mineral kingdom prepares the oxides, for the use of organised bodies.

The earth, we have said, supports plants, and gives them a fulcrum or point of rest, which animals have not, because they do not require it. In order to support plants, the ground must be neither too compact nor too loose. They cannot grow upon a hard rock, nor in a moving sand: their roots cannot penetrate the first, nor take firm hold on the latter.

EMILY.

Besides, a plant would find no food to nourish it on a barren rock.

MRS. B.

That most patient and persevering of agriculturists, Nature, teaches us how to prepare a soil, even on the hard rock, or the sterile lava of a volcano: she commences her operations on these obdurate bodies by means of her elementary agents, air and water. If the rock be of a calcareous nature, the lime is gradually dissolved, a decomposition begins to be effected; and hence the origin of a soil. If the rock be silicious the operation is more difficult; but Nature, unrestricted by time, finally accomplishes her object. On this shadow of a soil, a vegetation, almost imperceptible from its minuteness, begins to exist: the invisible seeds of lichens, which are ever floating in the air, there find an asylum. Minute as these seeds are, they are furnished with admirable means to attach themselves to hard bodies.

CAROLINE.

What a careful provision Nature has made for the most insignificant of her vegetable kingdom!

MRS. B.

These seeds, once attached to the rock, find sufficient nourishment in the moisture they have absorbed from the atmosphere during their aerial flight to enable them to germinate, but not sufficient to bring them to perfection, and enable them to produce seeds to continue their species; but when they perish, a new race rises, phœnix-like, from their remains.

EMILY.

This is indeed a new mode of raising plants!

MRS. B.

It is these remains, mixed up with some of the crumbled particles of the rock, which constitute the first bed of earthy soil, in which the seeds of more robust lichens and mosses sow themselves, and find nourishment; thus a variety of plants, annually increasing in strength and vigour, rise up in succession, till the dry rock becomes covered with verdure, and ultimately clothed with trees You see, therefore, Caroline, that you have no more reason to despise the humble plants in whose remains a soil originates, than to underrate the germ of a shoot which may produce a stately oak.

There is another process which Nature frequently employs to clothe a barren rock: whereever there are fissures the rain insinuates itself, and by freezing in winter often splits the rock, or at least widens the crevices; it is in these humid recesses, where the water has crumbled the rock, that seeds bury themselves, and vegetation commences.

EMILY.

In attempting, then, to cultivate a barren soil, we should follow those lessons which Nature points out, and scatter seed in such crevices, which, if they did not arrive at complete maturity, would, by their remains at least, help to prepare a soil.

MRS. B.

This is often done; for instance, in the fissures of the lava of Mount Etna. Indian fig or prickly pear has been sown; the roots of which insinuate themselves into every little cavity, and help to split the block. This plant produces a great quantity of fruit, but its most important recommendation is, that of forming a soil for future vegetation; but to proceed to soils of a less obdurate nature.

A stiff argillaceous soil is difficult to cultivate, on account of the resistance it opposes to the penetration of roots. This description of soil attracts moisture, and is so retentive of water as to be seldom dry, unless during the heat of summer, when it splits; and it is in the crevices thus formed that vegetation commences. Such a soil requires frequent ploughing, in order to break down and pulverise the clods, when practicable, earth of a lighter nature should be mixed with it. Plants having large roots will not, in general, succeed in a soil of this description, as they will not be able to penetrate it.

EMILY.

Yet plants with small delicate roots will have still less strength to do so.

MRS. B.

True; choice should therefore be made of plants which have slender, but firm and strong roots: those whose roots are of a dry nature are best adapted to hard, impenetrable soils.

A sandy soil offers difficulties of an opposite description. If the sand be mixed with calcareous matter, these are more easily overcome; for a portion of the lime is dissolved by rain, and its solution gives some degree of stability to the soil: but if the sand be almost entirely silicious, like that of the sea, the evil is well nigh insuperable; for this species of sand is insoluble, and nothing can change its nature. Hence the impracticability of cultivating the sandy deserts of Arabia, Africa, and various other parts of the world.

EMILY.

Yet, if such deserts existed in Europe, do you not think that means would be discovered to overcome the difficulty?

MRS. B.

I doubt it. The wind which blows without restraint over these unsheltered and unstable plains tears up the roots of every tree that is planted.

EMILY.

But small low shrubs would offer but little resistance to the wind.

MRS. B.

These would soon be buried by the whirlwinds of sand. The plants most likely to succeed would be such as are of moderate stature, with spreading roots to fix them in the soil.

When sands are of small extent they may be improved by mixing clay with them; and the first crop should be raised solely with the view of ameliorating the soil.

There are three species of sandy soil: that which forms the banks of rivers; that which composes those extensive plains called steppes; and that which forms sand-hills on the sea-shore. On the borders of rivers, stakes of willow and of alder may be planted with advantage; being abundantly watered, they soon shoot out roots and branches, which grow rapidly. Then, if with the stroke of a hatchet these branches be lopped, so as to make them trail upon the ground, without being completely separated from the stem, they will soon be covered with the loose soil and will strike fresh roots: these numerous roots shooting out in every direction, are interwoven, and form a species of network, which sustains and gives stability to the soil.

In order to bring steppes into culture, which are not so well supplied with water, the first plants raised must have roots which pierce deep into the earth, so that they may find water beneath the sandy soil. The culture of madder has been successfully employed in the neighbourhood of *Haguenot*, as the precursor of general agriculture. It cannot be too often repeated, that when you aim at bringing bad soils into culture, the first produce must be sacrificed for the benefit of the land, with a view to improve it for future harvests.

CAROLINE.

So far as can be judged from the abundance and magnitude of the crops, Belgium appears to be one of the countries in which agriculture is carried to the greatest degree of perfection.

MRS. B.

The beauty of the produce is no bad criterion of the advancement in the art, especially in Belgium, where Nature has done little for the husbandman; but the Belgic peasantry are nearly as well versed in agriculture as the learned of other countries. Their soil is in a great measure the work of art, man having taken possession of it before Nature had completed its formation.

CAROLINE.

Nor does it appear that the art of man has yet finished it; for though the cultivated parts yield such rich crops, an extensive sandy desert, called the Campine, still remains on the confines of Belgium and Holland.

MRS. B.

True; but cultivation advances with gigantic strides across the arid waste. The mode by which the husbandman commences the process of fertilisation in these sterile plains is by sowing Genet or Broom, which grows up in bushes, the roots of which confine the soil, and give it sufficient stability to enable him to sow pines with advantage. These are followed by Acacias; the branching roots of which, stretching out in various directions and interwoven together, sustain the soil as it were in osier baskets. But it is not until this succession of forests have flourished and decayed that the soil, enriched by their remains, becomes fit for general culture.

EMILY.

This, then, is the work of a long course of years?

MRS. B.

Certainly; but still the formation of the soil is rapid, in comparison of what it would have been, if left to be completed by the gradual agency of Nature, who is enabled to prolong her operations beyond the period of our transitory existence, and is therefore less impatient to accomplish her task. We are justified, however, in taking it out of her hands, if we can produce the effect more rapidly.

The sand-hills, which are, in many places, formed on the sea-coast, owe their origin to sand thrown up by the high tide, and which, abandoned by the receding waters, dries, and is carried by the wind farther inland, and out of the reach of successive tides. The sand-hills formed in the vicinity of Bourdeaux formerly threatened the destruction of the adjacent country: it was calculated that no less than seventeen villages would be overwhelmed by them in the course of a century; when M. Bremontier was so fortunate as to discover a means of averting this danger. Observing that sand thus thrown up was not devoid of moisture, he scattered over it the seeds of broom and of maritime pine; and, in order to prevent their being swept away by the wind, he covered them with brambles and branches of underwood. The seed sprouted; the broom first rose above ground, and some time after the young pines appeared: the latter, however, made but little progress, seeming to be choked by the rapid growth of the broom; yet in the course of a few years the pines gained the ascendency, and drove their antagonists from the field; or rather, I should say, like true cannibals, after destroying the enemy, they fed upon their remains.

In the course of time it became necessary to thin this vigorous forest of pines; and their branches served to shelter the seed scattered on neighbouring sand-hills.

CAROLINE.

I recollect reading in Withering, that the Arundo arenaria, or sea-matweed, which grows only, on the very driest sand on the sea-shore, prevents the wind from dispersing the sand over the adjoining fields. The Dutch have very probably known and profited by this fact.

MRS. B.

Hitherto we have directed our attention rather to the formation of new soils than to the improvement of old ones: yet the latter is the point of most importance in agriculture; for we are much more frequently called upon to ameliorate the land already under tillage than to prepare a soil, on land which has not yet yielded any produce.

Soils may be improved by a variety of different processes: by tillage, by amendments, by manure, and by rotation of crops.

These follow each other in natural succession. Man first begins by cultivating the earth; he next endeavours to ameliorate the soil, in order to render it more propitious to the produce he wishes to raise. After having yielded a certain number of crops, he observes that the earth is exhausted of its nutritive principles, and that the crops are poor and meagre. He finds the means of renovating these principles by manuring the land; and, when manure falls short, he discovers that a judicious system of cropping answers, in a great measure the same purpose.

The principal object of tillage is to break and crumble the earth, in order that the roots of young plants may easily penetrate into it; to expose every part of the soil successively to the action of the air, so that such of the earths or metals, as are destined to be converted into salts by the action of the oxygen of the atmosphere may be brought into contact with it, as well as such remains of organic bodies as can be dissolved only by oxygen. Various implements of husbandry are employed for this purpose; and it is a continual object of dispute, between the agriculturists of different countries, which answers the purpose best.

EMILY.

But is not the plough the instrument universally used in all civilised countries?

MRS. B.

Most commonly; but the plough itself is of various descriptions, and you have observed the peasantry of Tuscany frequently employing the spade, an implement which we reserve for garden culture. They are, indeed, bound by the tenure on which they hold their land to dig it up every third year. The spade undoubtedly performs the operation of turning and subdividing the earth more completely than the plough, but at a much greater expense of labour; and it is an instrument adapted only to light and homogeneous soils, for if the earth be of unequal tenacity, or interspersed with stones, it cannot be used. The pickaxe may in those cases be substituted, as it is a pointed instrument, which more easily penetrates. This implement is also of various descriptions: it has a single or a double prong, which is broader or sharper, and forms a greater or lesser angle with the handle, according to the nature of the soil.

With the spade the labourer works backward, and throws up the earth before him; with the pickaxe he goes forward, and draws the earth towards him.

CAROLINE.

The hoe, that very useful instrument for weeding or lightly raising the earth, is also used like the pickaxe. But what is the reason that the form of these instruments vary so much, in different countries?

MRS. B.

Sometimes from improved models being adopted in one country, which another, through ignorance or prejudice, will not follow; and perhaps, more frequently, from the different nature of the soil. The spade or the hoe must be light or heavy, broader or more pointed, according as the soil is loose or stiff; for the heavier or more tenacious the earth, the less quantity can be raised at one stroke. But the most important of all implements of husbandry is doubtless the plough: it has been celebrated since the times of Moses and of Homer;

and it is the form of this instrument which has produced the greatest contention amongst agriculturists. The plough may be considered as a sort of pickaxe, drawn by animals through the soil. In northern climates husbandmen are great partisans of deep ploughing; in southern countries they are no less staunch advocates for light or more superficial ploughing; and they are, perhaps, each equally right in approving their own mode, and wrong in blaming that of their opponents, for the different species of plough are adapted to the soil of their respective countries.

In high latitudes, where there is much moisture and but little heat, it is necessary to turn over the earth more completely, in order to dry and pulverise it, especially when of an argillaceous nature, which is very common in northern countries. In more southern climes, where heat and drought prevail, it is better to plough more lightly, the soil being frequently of a sandy nature, not retentive of water.

EMILY.

But supposing the soil to consist of two layers, the one of sand, the other of clay, the plough should, I suppose, go deep enough to mix them together.

MRS. B.

No doubt their union produces as excellent a soil as their separation makes a bad one. It signifies not which is uppermost before ploughing;

the more they are mixed and incorporated together the better.

CAROLINE.

But supposing there should be a good, rich, vegetable soil on the surface, and layers of sterile ground beneath?

MRS. B.

Then a light plough should be used, and as much care taken to prevent the mixture of the two, as to effect it in the former case; in short, attention must always be paid to the nature and locality of the soil. Ploughing must vary, also, according to the nature of the produce to be raised. Lucern, which shoots out roots four or five feet in length, requires deeper ploughing than corn, whose roots are very superficial. Six or eight inches is a sufficient depth for grain in general.

EMILY.

When new land is first broken up, to bring it under tillage, it will, I suppose, require deep ploughing to pulverise the hard earth.

MRS. B.

That also depends on the nature of the soil. In America, where fine rich vegetable soil is daily brought into cultivation, nothing more is required than to scratch the earth with a plough, and scatter the seed, in order to produce an abundant harvest. But in England, and all countries which have long been cultivated, the good soil is already fully employed; and if any new land is ploughed up, it is of a very inferior description, and it is necessary not only to plough it deeply but repeatedly, and to manure it, before it will yield a crop. The operation of bringing grass land into tillage is on the Continent frequently performed by a pickaxe with a double prong, which breaks the earth more completely than the plough.

Another point to be considered in tillage is the quantity of manure to be spread upon the land. If this fall short, and the ploughing has been deep, the nutritive particles may filter down lower than the roots can go in search of them.

The more tenacious and compact the soil is, the closer the furrows must be, and the narrower the ridges of earth turned up, in order more effectually to pulverise it, and afford channels for the water to run off. When the soil is light, broader ridges and more distant furrows suffice: it is even sometimes necessary to beat down the earth, after having ploughed or dug it, in order to render it more compact, especially in nurseries of young trees, whose roots, in a loose soil, are liable to be torn up by the wind.

Deep furrows, or trenches, are very useful where the ground is sloping, either to draw off or retain the water as required. If the soil be too moist, the furrows should be made longitudinally, that is to say, from the top to the bottom of the acclivity: they will then answer the purpose of conduits to carry off the water. If, on the contrary, the soil be dry, the furrows should be made transversely, and the ridges of earth will act as parapet walls to retain the water.

It is very necessary also, to pay attention to the period of ploughing: it can be done neither in a wet season nor during a hard frost, nor in very dry weather; but as you have the whole season before you, from the reaping of one harvest to the sowing for another, it is not difficult to choose a period of appropriate weather, unless it be in some strong clays, upon which a horse cannot be suffered to tread during the winter. If the ground be intended to lie fallow, the best use which can be made of the repose allowed it, is to plough it in autumn, and again in the spring; but if it is to be sown, the sooner it is ploughed after harvest the better, in order to bury the straw or other remains of the preceding crop, which will enrich the soil, and also prevent the further growth of weeds.

In hot countries the land cannot be ploughed in summer, on account of its dryness; besides, it would afford the means of evaporating the small remains of moisture of the newly-turned up earth, and that at a period, when it has its most important functions to perform — those of softening and dissolving the hardest and most insoluble particles, which cannot be done unless the temperature of the water be tepid.

CAROLINE.

But how can it be rendered tepid without exposing it to the sun?—And in that case it will be evaporated.

MRS. B.

Exposure is not necessary: the heat of the atmosphere gradually penetrates the soil, and the water diffused in it acquires the same elevation of temperature. Farmers conceive that the soil is injured by the action of ploughing in summer; but the injury proceeds from impeding the solutions requisite for the following crop. In northern climates, where evaporation is less active, ploughing is not so objectionable in summer.

CONVERSATION XIII.

THE ACTION OF SOIL ON PLANTS CONTINUED. — ON THE IMPROVEMENT OF SOIL.

MRS. B.

WE may now proceed to examine the various modes of improving the soil by mineralogical processes. The first and most simple of these is to clear it of stones; when stones are injurious to cultivation.

CAROLINE.

But are not stones always injurious? For of whatever materials they may be composed, they are such hard insoluble bodies that vegetables can acquire no nourishment from them.

MRS. B.

True; but stones often perform a very useful mechanical part in agriculture. They render a clay soil less tenacious by separating its parts, and thus leave room for water to drain off: they form, as it were, so many natural irregular conduits; and, if you take them away, you must employ them for the construction of artificial conduits to effect the same purpose.

CAROLINE.

This may be the case with stones buried in the earth, but those lying on its surface must surely be prejudicial to vegetation?

MRS. B.

Generally they are so, but not universally: in some hot countries grass cannot grow, excepting under shelter of loose flat stones. I have seen pastures of this description on the plains of Crace, near Arles. They exhibit the singular spectacle of flocks of sheep feeding on dry stones, as the grass which grows beneath them is not visible; but the sheep find a tender, if not abundant pasture, by turning up these stones, or nibbling beneath them; while the pebbles, thus overturned, afford shelter to the adjacent blades which are just sprouting, and would be burnt up without such protection.

In some instances the ground in which fruit-trees are planted, have been paved with stones, in order to retain the moisture beneath by preventing evaporation.

In cold countries stones are sometimes considered advantageous as communicators of heat, those of a dark colour especially. They act on the surface of the earth both as reflectors and radiators of heat; and are frequently placed round the stems of plants in a vineyard, in order to give them additional heat. It must be allowed, however, that those occasional uses to which stones are applied in husbandry, are to be considered rather as exceptions to the general rule, and that stones may be looked upon in most cases as either useless or pernicious.

The improvement of soil by the admixture of foreign ingredients, *amendement*, is one of the most important operations of agriculture.

If the soil be too stiff from excess of clay, it will be improved by sand; and if too loose from excess of sand, it will be improved by clay: but when sand is mixed with argillaceous soil, the latter must be broken and pulverised, which may be effected by exposing it to the frost, and afterwards drying it. Marl is a natural compound earth, used with great success in the amelioration of soils: it consists of a mixture of clay and calcareous earth or lime in various proportions.

Argillaceous marl, which contains more clay than lime, is advantageous for a dry sandy soil; while calcareous marl, in which the lime predominates, is suited to an argillaceous soil. The great advantage of marl is, that it dilates, cracks, and is reduced to powder by exposure to moisture and air. Marl in masses would be totally useless on the ground; yet it is necessary to begin by laying it on the ground in heaps, for the more it is heaped the more it dilates, splits, and crumbles to dust; in which state it is fit to spread upon the ground.

Marl is sometimes intermixed with other earths; sometimes formed into a compost with manure before it is laid on the soil: it should be applied sparingly at a time, and renewed frequently. Its advantages are manifest: it subdivides the soil and accelerates decomposition, its calcareous particles disorganising all animal or vegetable bodies, by resolving them into those simple elements in which state they combine with oxygen; it facilitates this union: hence, though not itself of a nutritious nature, it promotes the nourishment of plants by preparing their food. The best period for marling ground is the autumn.

Lime is also an excellent *amendement*. It is procured from limestone by exposing it to the heat of the kiln, which evaporates the water and carbonic acid with which lime is always found combined in nature, and renders it *quick*, as it is commonly called; that is to say, of a caustic burning nature, having such an avidity for water and carbonic acid, from which it has been forcibly separated, that it seizes upon these bodies, wherever they are to be met with, and disorganises the compounds in which they are contained in order to combine with them.

EMILY.

But since lime is of so destructive a nature, I should have thought that it would have been necessary to add, instead of subtracting, water and carbonic acid, in order to soften its caustic properties, which seem calculated rather to destroy than promote vegetation.

MRS. B.

Were quick-lime applied immediately to plants, it is true that it would prove a poison to them; but, when spread upon the earth, it rapidly attracts water and carbonic acid from the atmosphere, and it is only when thus modified that it promotes vegetation.

EMILY.

Then why force from it, in the kiln, those very ingredients which must be restored to it so soon afterwards?

MRS. B.

In its natural state of limestone it is of too hard and compact a nature to be diffused in the soil; and even quick-lime would be too solid, were it not, that through its combination with water and carbonic acid from the atmosphere, it splits and crumbles to powder.

I believe the experiment of pounding and grinding limestone to powder, was tried in Scotland, in order to save the expense of burning it into quick-lime, but not found to be efficacious like the powder of slacked lime.

Lime is particularly adapted to poor cold soils, such as those of marshes, which have not energy to dissolve organic bodies. The quantity to be used must be proportioned to the manure which is laid upon the ground; for the more organic matter there is to be dissolved, the greater will be the quantity of lime required for that purpose. To mix lime with peat-earth, is said to have an immediate and most beneficial effect, and that many bogs, having been previously drained, have been converted into fertile land.

The lime procured from fossil-shells is highly esteemed by agriculturists: its pre-eminence results, probably, from its retaining some vestiges of organic remains of the animals who once inhabited these receptacles.

CAROLINE.

The shells of living animals must then be still more valuable for this purpose?

MRS. B.

They are neither so readily, or so abundantly obtained; large strata of fossil-shells are to be found in some soils, whilst of living shells you could procure at most, the refuse of the fish-market.

Ashes are very beneficial to the soil: they differ much in their composition, according to the nature of the body from the combustion of which they result, but their general ingredients are potash, silex, and calcareous earth. They attract moisture from the atmosphere, and thus accelerate vegetation.

Sulphate of lime, commonly called gypsum, is an excellent *amendement*; but chemists are not agreed as to the manner in which it acts on vegetation. It is strewed over crops when the leaves are in full vigour, towards the latter end of April or the beginning of May, and it should not be laid on more than once in the year. Clover and saintfoin contain gypsum in their stems to a considerable amount; and when soils are said to be tired of those plants, it is probable that they are no longer able to supply this necessary ingredient. It is on those crops that gypsum is found to be most efficacious.

Having now made you acquainted with the various modes of improving the soil, we are next to consider which are the best means of supplying plants with food.

CAROLINE.

All natural soils, with the exception, perhaps, of burning sands, or arid rocks, must contain nourishment for plants; otherwise they would not grow spontaneously as they do in wild, uncultivated countries, which often abound with forests and rich pastures.

MRS. B.

True; but though this supply be sufficient for a natural state of vegetation, when the land is forced, as it were, by agriculture to yield food for man, a greater produce must be obtained; and we cannot raise those rich and numerous crops, so necessary to the existence of a civilised country, without affording the vegetable creation an artificial supply of nourishment: for it is an axiom, no less true in the vegetable than in the animal kingdom, that food must be proportioned to the population, in order to maintain it. The mode which art has devised to increase the quantity of food for plants is to spread manure on the soil. Manure consists of the remains of organised bodies of every description, whether animal or vegetable, in a state of decomposition; that is to say, resolving itself into those primitive elements which can re-enter into the vegetable system.

CAROLINE.

The preparation of food for plants is then precisely the inverse of that for animals, or at least for animals of the human species. Our culinary art consists in mixing and combining together a variety of ingredients to gratify the palate; whilst bodies must be decomposed and resolved into their simplest elements to suit the vegetable taste. — And how is this process performed?

MRS. B.

I have explained it to you in our Conversations on Chemistry: it is by the last stage of fermentation —*putrefaction*. Loathsome as this term may appear to you, yet, when you consider it as the means which Nature employs to renovate existence, and continue the circle of creation, you will view it with admiration rather than with disgust.

EMILY.

It is very true; the operations of Nature, when

philosophically contemplated, are always admirable: those elementary substances, which, in their simple state would be disagreeable to us, by passing into the vegetable system, are converted into the most palatable and nutritious food. When in the resources of Nature we discover such evident proofs of the goodness of the Creator, the philosopher may well exclaim with the poet:—

> " These are thy glorious works, Parent of good,
> Almighty! Thine this universal frame
> Thus wondrous fair; thyself how wondrous then!
> Unspeakable! who sitst above these heavens
> To us invisible, or dimly seen
> In these thy lowest works; yet these declare
> Thy goodness beyond thought, and power divine."

MRS. B.

This beautiful burst of praise, into which Adam breaks out on the creation at large, is no less applicable to the wisdom and prudence displayed in the arrangement and distribution of its minutest parts.

The principal result of the decomposition, whether of animal or vegetable matter, is carbonic acid; and in this state carbon, which we have called the daily bread of plants, finds entrance at their roots.

CAROLINE.

But it is not enough to introduce carbonic acid at one extremity of the plant: you must get rid of the oxygen at the other extremity before the plant can feed upon its daily bread.

EMILY.

This, you may recollect, is performed by the leaves when exposed to light and air.

MRS. B.

Manure acts on dry soils also as *amendement* by retaining moisture. Manure which has not completely undergone the process of fermentation, so that the straw is not yet wholly decomposed, is best adapted to strong compact soils: the tubular remnants of straw answer the purpose of so many little props to support the earth, and afford a passage for the air, thus rendering the soil lighter; besides, the completion of the fermentation taking place after the manure is buried in the soil, has the advantage of raising its temperature.

EMILY.

Since the putrid fermentation reduces every animal and vegetable substance into its primitive elements, there are none, I suppose, which may not be converted into manure?

MRS. B.

None; but some bodies are more readily decomposed than others: it is from domestic animals that the best manure is obtained. In maritime districts, fish, when sufficiently abundant, is sometimes used to manure the land. They are easily decomposed, and afford a considerable quantity of nourishment. Even such hard substances as horn, hair, feathers, and bones, are all resolvable into their primitive elements, and make excellent manure; but, owing to their dry nature, require a longer period for their decomposition. Such substances are calculated not for annual harvests, but to fructify the soil for a produce of much longer duration, such as that of olive-trees and of vineyards.

Vegetable manure does not always undergo fermentation previous to being buried in the soil: green crops, such as lupins and buck-wheat, are sometimes ploughed in, and thus buried for the sole purpose of enriching the soil. A green crop contains a considerable quantity of water; and the plants, when buried, serve to lighten the soil previous to decomposition, and subsequently to enrich it with food for the following crop. This species of living manure is particularly calculated for hot countries, on account of the abundance of moisture it incorporates with the soil.

EMILY.

I have seen sea-weeds used as manure, which has at least the advantage of being a gratuitous crop.

MRS. B.

Gratuitous in some respects, but requiring a difficult and laborious carriage. The Isle of Thanet owes its reputation in a great measure to the power of procuring this manure. And the sea-salt they contain is also favourable to vegetation. Straw is an excellent ingredient for manure; but it requires being mixed with animal manure, or stratified with earthy matter. Bark and sawdust are occasionally used for manure: they should, however, with greater propriety be considered as materials for improving the soil, as they afford but little nourishment.

The grain which produces oil, such as linseed, rapeseed, &c., makes excellent manure after the oil has been expressed: in this state it is called oil-cake, and its unctuous qualities serve to accelerate decomposition; but in England it sells at such a price as to make it a doubtful speculation even to feed cattle with. It would be too expensive to be used as manure.

EMILY.

Pray, would not soot make very good manure? It is almost pure carbon, in so highly a pulverised state, as must render it fit to enter into the vegetable system.

MRS. B.

You forget that it is first necessary to combine it with oxygen; and this is a work of time. Soot has, however, an immediate beneficial effect, though not a very permanent one: it is used in large quantities in Hertfordshire, both for grain and pasture, and is strewed on the land in March and April, for the crop of the same year.

CAROLINE.

I recollect having observed that the environs of the spots, where charcoal has been prepared in the mountains, are absolutely destitute of vegetation, although strewed with charcoal-powder.

MRS. B.

But were you to visit these same spots some few years afterwards, you would find vegetation more flourishing, more vigorous, and especially greener than elsewhere, because the charcoal-powder will have gradually combined with the oxygen of the atmosphere, and thus vegetation be luxuriously supplied with its favourite food, carbonic acid.

The most common manure consists of a mixture of animal and vegetable materials; and this, again, is frequently stratified with mineral substances, such as mud from the streets, dust from the roads, or earth of different descriptions, the whole forming a rich compost. Mud from the beds of rivers, when it can be obtained, is a very valuable ingredient for such a compound, as it abounds with organic remains of fish, shells, reptiles, and rotten plants. Often, however, before being laid upon land, it requires being well turned up and exposed to the air for some time.

CAROLINE.

It is to be regretted that such precious relics should, in general, be lost by being carried by the current of the river into the sea; but the slime of ponds and all stagnant waters must make very rich manure.

MRS. B.

Yes; they may be considered as storehouses of materials, ready to return into the vegetable system.

The elevation of temperature produced by the completion of fermentation of the manure, after it is mixed with the soil, is but inconsiderable, excepting in garden culture, where, accumulated in hotbeds, it often produces a temperature equal to that kept up in a hothouse.

Short manure, that is to say, that which is thoroughly decomposed, and in which the water and other evaporable parts have in a great measure disappeared, contains a considerable quantity of carbon.

Long manure, in which state fermentation is but little advanced, contains a greater proportion of water: the first is, therefore, best adapted to moist, the latter to dry, soils.

EMILY.

But if the fermentation be completed previous to mixing the manure with the soil, are there not many volatile products which escape into the atmosphere, and which might, if buried in the earth, have promoted vegetation?

MRS. B.

No doubt. It is incalculable how many valuable materials are "wasted on the desert air" which should have given vigour to vegetation; others are dissolved by moisture, and drained off by rain; but these liquefactions are generally collected and turned to account. To prevent, as far as possible, such losses, dunghills should be sheltered from the atmosphere by sheds: these should, however, remain open on the sides, as air in a moderate quantity is required to promote fermentation.

EMILY.

In which state do you consider it most advantageous to bury manure in the soil: when the fermentation is only partially, or when it is completely effected?

CAROLINE.

I should suppose in the former state, in order to prevent the loss by evaporation. When the fermentation goes on in the soil, the elastic as well as the liquid and solid parts are retained; then the act of fermentation raises the temperature.

MRS. B.

One state is better for some species of crops, and the opposite for others. The only disadvantage attached to long manure is, that it requires a greater length of time to convert it into nourishment for plants.

Short manure is a meal already cooked, and ready for the crop to feed on; if, therefore, the crop requires very prompt sustenance, the former must be used; if not, the latter is in every respect preferable: it is particularly adapted to stiff soils, the straw, previous to its decomposition, rendering it lighter.

EMILY.

It is evident that it must be advantageous to bury either description of manure as soon as it is spread on the soil, to prevent loss by evaporation; but how deep should it be laid in the soil?

MRS. B.

That depends upon the nature of the culture; for the manure should be as much as possible within reach of the roots. For this purpose, it should not be buried quite so deep as the extremity of the roots; for, in proportion as it is dissolved and liquefied, it will naturally descend. Due allowance must be made for this; for, if any part subside below the roots of the plants, it is utterly lost, at least for that crop.

EMILY.

It is then, I suppose, better to manure the land in the spring than in autumn, lest the winter-rains should dissolve it too much, and endanger its sinking below the roots of the crop.

MRS. B.

That is the prevailing opinion of agriculturists. With regard to the quantity of manure, it is a commodity so scarce, that it is not likely to be employed in excess. This occurs, however, sometimes in garden culture, and it produces a strong and disagreeable flavour in the vegetables; even the cows avoid the strong coarse grass which grows on spots they have manured too abundantly. But the stock of manure is generally so limited, that it has been the study of agriculturists to discover some means of compensation for a deficiency, rather than to apprehend danger from excess. This compensation has been found in a judicious system of crops; but it is too late to enter upon a new subject to-day, and one of so extensive a nature well deserves to have a morning dedicated to its consideration.

CONVERSATION XIV.

THE ACTION OF SOIL ON PLANTS CONTINUED. — ROTATION OF CROPS.

MRS. B.

It has long been observed, that two successive harvests of the same species of plants, or even of plants of the same family, do not succeed: the second appears to degenerate, as if the first had been injurious to it.

CAROLINE.

The first crop had no doubt exhausted the soil of nutriment.

MRS. B.

In the infancy of agriculture, when land was plentiful, because inhabitants were scarce, this was easily remedied by cultivating only a certain portion of land, and, after having exhausted it, transferring the cultivation to another part, and thus successively bringing new land into tillage, till, after a series of years, they return to the spot which had been previously cultivated. This mode, called *ecobuage*, was first introduced by the Celts, and may still be traced among some of their descendants in Brittany. They usually commenced their operations by burning the natural produce of the soil before they ploughed it. If the soil was stiff and argillaceous, the ashes resulting from this combustion seemed to ameliorate it, by increasing the stock of carbon, of sand, and of salts; but if light, such a proceeding was not judicious.

The system of fallows, which we derive from the Romans, is an improvement on that of the Celts; the soil is allowed only one year's repose occasionally, and during that season it is repeatedly turned over by the plough; every part is thus exposed to the atmosphere, whence it absorbs oxygen, and the weeds, being buried by the plough, serve to enrich, instead of exhausting, the land.

The system of *assolemens* we owe to those excellent farmers the Belgians. It is of two descriptions: the first consists in the judicious cultivation of such a succession of crops, that they shall derive benefit instead of injury from each other; the second is that of raising two crops simultaneously, which shall mutually benefit each other. As we have no precise term to express these processes, I shall take the liberty of using the French word *assolement*.

Those of the first description, which our farmers denominate a rotation or course of cropping, is particularly adapted to northern climates.

EMILY.

And of what description are the crops which ought to be cultivated in rotation?

MRS. B.

Before this point can be determined, we must endeavour to sol ve the problem why one crop is prejudicial to another of the same family; why two sorts of grain cannot be raised in succession without the latter degenerating; whilst leguminous plants succeeding a crop of grain are improved by it.

EMILY.

I thought that the first crop, especially if one of grain, which requires so much nourishment, would injure the soil for any succeeding crop by exhausting it of nutritive particles.

CAROLINE.

But since it appears that leguminous plants can follow grain with advantage, it seems evident that these two crops must feed upon different kinds of nourishment, and thus the one will not interfere with the other.

EMILY.

You forget, Caroline, that plants cannot select their food, but suck up whatever comes within reach of their roots, and is sufficiently minute to find entrance there. All plants must, therefore, feed upon the same nutritive particles.

CAROLINE.

Then I will suggest another explanation. Perhaps the roots of plants which succeed each other without injury may be of different lengths, and one crop seek its nourishment near the surface, while the other penetrates deeper into the soil; thus they would both be fed without interfering with each other.

EMILY.

But you do not consider, Caroline, that the plough overturns your theory, in overturning the soil; for it brings the lower part to the surface, and mixes the whole so well together that the nutritive particles must be pretty equally diffused throughout.

MRS. B.

Your observation is very just; and we find that clover, which has very superficial roots, will not thrive after lucern, whose roots are very long. This theory, however, is applicable to simultaneous crops whose roots are of different lengths. Another theory has been suggested, which is, that the fall of the leaf of the first crop fertilises the earth for a second: this is undoubtedly true to a certain extent; but the foliage can fertilise the earth only by being converted into manure, which would equally afford nourishment for a second crop of the same nature.

CAROLINE.

Nay; one would even suppose that a green crop, ploughed into the soil, would afford more appropriate food for a second crop of the same description than for one of a different family; that the leaves of straw would yield the best nutriment for a future crop of corn, and of grass for that of grasses; whilst the fact, you say, is exactly the reverse.

MRS. B.

The theory which M. De Candolle is most inclined to favour, if indeed he is not its author, is the following. A plant, being under the necessity of absorbing whatever presents itself to its roots, necessarily sucks up some particles which are not adapted to its nourishment, and in consequence,— after having elaborated the sap in its leaves, and re-conducted it downwards through all its organs, each of which takes in the nourishment it requires; after having extracted from it the various peculiar juices, and, in a word, turned it in every possible way to account, — finds itself encumbered with a certain residue, consisting of the particles it had unavoidably absorbed, and which were not adapted to its nourishment: these particles, having passed through the system without alteration, are exuded by the roots which had absorbed them, and thus return into the soil, which they deteriorate for a following crop of the same species of plant, but improve and fructify for one of another family; thus affording an admirable proof of the wise economy of Nature, in multiplying her vegetable produce by feeding different plants with different substances, and enabling beings, incapable of distinguishing or selecting their food, to obtain that which is appropriate to them.

EMILY.

It is, indeed, admirable! Then, though the roots of plants can make no choice, their organs are in some measure capable of selecting, since they reject, and will not elaborate, substances which are not adapted to the nourishment of the plant.

MRS. B.

If we cannot exactly allow them the nice discrimination of the chemist, we must at least suppose their laboratory to be so arranged as to act only on bodies congenial to the plant.

CAROLINE.

And the rejected substances, which would be poison to one family of plants, when transfused into the soil, is greedily devoured by a succeeding crop of a different family.

EMILY.

Yet, Mrs. B., there is land in the Vale of Glastonbury, in Somersetshire, which is celebrated for growing wheat for many years together without any manure; and I have heard that in the neighbourhood of the Carron iron-works, in Scotland, wheat has been raised above thirty years, without injury either to the crops or the soil.

MRS. B.

Those soils must not only abound with vegetable nourishment, but the land be particularly well adapted to growing wheat; consequently, the roots would have little or nothing to exude, and successive crops of wheat might be raised so long as the land was not exhausted. This explanation would reconcile your difficulty to the theory of exudations; but, interesting and plausible as this theory is, it requires the confirmation of facts to rest on a solid foundation: few experiments have yet been made relative to it. Mr. Brookman has raised some plants in sand, and ascertained that they exuded by the roots small drops during the night, which there is reason to suppose was the object in research; but experiment has not yet been pushed far enough fully to verify it.

CAROLINE.

It appears to me to explain the theory of *assolement* so well, Mrs. B., that I feel strongly inclined to put my faith in it. How perfectly it accounts for the advantages derived from cultivating leguminous and graminiferous crops in succession, the exudations of the one being exactly the nutriment which the other requires.

MRS. B.

I am so much inclined to agree with you in opinion, that I shall venture to draw conclusions from it, as if the theory were established; cautioning you, however, to bear in mind, that until it has been further investigated, it must be considered as little more than hypothetical.

You must besides remember that it is manure which affords the grand store of provisions equally good for plants of every description. If, in addition to the exudations of leguminous plants, you plough in the crop itself, the succeeding crop of corn will be considerably improved.

EMILY.

Supposing you were to plough in a crop of young rye, could you not the following year sow wheat with advantage? for the rye would have had but little time to deteriorate the soil by its exudations, and afford much manure by the fermentation of its own substance.

MRS. B.

I do not know that the trial has ever been made;

but it would certainly succeed better than if you were to reap the rye when ripe, and afterwards sow wheat: for, in this case, the rye would have given to the soil the whole of its exudation, and little or none of its own substance.

CAROLINE.

May it not be objected to this theory that Nature does not raise plants of different families in succession? The seeds of the parent-plant fall to the ground annually, and produce other individuals of the same species, and on the same spot, for centuries; and yet that spot must be vitiated for such plants by the long series of exudations of their progenitors.

MRS. B.

If Nature does not usually employ successive, she does simultaneous *assolements*. In her spontaneous forests she raises such a prodigious variety of trees and shrubs, and in her meadows such a multiplicity of herbs and grasses, that the different plants mutually supply each other with exudations.

EMILY.

Besides, where Nature acts without restraint, she enriches the soil not only by the annual fall of the leaf, but, in the course of time, the whole plant, whether grass, shrub, or tree, returns to the soil, to repay the nourishment it had received during its life.

MRS. B.

The soil can never be impoverished by natural vegetation: that of forests, where man does not cut down and carry away the trees, is not more exhausted of nutriment at the present day than it was a thousand years ago.

Those magnificent forests which covered the face of the greater part of America, when it was first known to us, never had any other manure than the remains which its vegetable and animal productions afforded; nor can a better be supplied. And we in some respects copy Nature when we prepare the soil for corn by ploughing in, a green crop of leguminous plants.

There is nothing which exhausts either a plant or the soil in which it grows, so much as the ripening of its seeds. No animal labours with greater effort to support its offspring than the poor plant to bring its seed to maturity: it pumps up sap with all its powers of suction; yet, if it has much seed to ripen, after having accomplished its task, it frequently perishes through exhaustion from the intensity of its efforts.

Perennial plants have but few, and but small grains to ripen, while those of annuals, are large and much more abundant; and it is this difference, perhaps, which constitutes the real distinction between these two classes of plants: the one, exhausted by its efforts, dies after ripening its seed; whilst the other, having a less laborious task to perform, lives through several successive years.

CAROLINE.

If that is the only distinction, an annual might live several years, were its seed prevented ripening.

MRS. B.

Instances of this sometimes occur in cold countries, such as Scotland. If the season has not afforded sufficient heat to ripen the corn, and that the following winter has not been so severe as to prove fatal to it, it will ripen the succeeding summer; and, indeed, whenever by any artificial means you prevent the ripening of the seed of an annual, it becomes perennial.

But to return to our subject. The succession of crops should be so arranged as to prevent as much as possible the growth of weeds: but what plant is it which deserves so opprobrious a title? for not one issues from the hands of its Creator which is not destined to act some useful part in its own sphere: either its exhalations purify the air; its exudations fructify the earth; its fruit supplies us with food or clothing; its blossoms regale our senses; and even its poisons minister to our diseases. What plant can we then denominate a weed? — The only blame which attaches to weeds is (as Dr. Johnson expresses it) being out of their place; and it is the business of the agriculturist so to fill up the place they would occupy, as to drive them out of the field. This cannot be more effectually accomplished than by the cultivation of artificial grasses, such as clover and lucern, which, when sown thick, produce a shade very prejudicial to the growth of weeds; if sown thin, so as to leave space, light, and air, it, on the contrary, encourages their growth.

There is nothing more favourable to the improvement of land than hoed crops, provided no immediate profit be expected from them, and that we are satisfied if they repay the expenses of cultivation; that is to say, the value of the seed, the hoeing, the ploughing, and the manure.

EMILY.

But why are not these crops sown so thick as to prevent the growth of weeds, and, consequently, the necessity of hoeing?

MRS. B.

These crops consist of plants whose roots require a greal deal of nourishment, such as peas, beans, turnips, potatoes, and carrots; and, if sown thick, the soil would not afford a sufficient supply.

EMILY.

Yet the weeds which spring up between these plants must rob the soil of a part of its nourishment.

MRS. B.

They do so, but only temporarily; for the dead weeds, after hoeing, return to the soil in the form of manure. The advantage of hoeing is not confined to the destruction of weeds: it loosens the earth so as to admit the air, turns it over, and heaps it around the roots of the plants cultivated.

As hoed crops stand in need of a great deal of manure, they should precede grain, which also requires manure to ripen the seed; and it is from the sale of grain, raised under these advantageous circumstances, that the cultivator will derive his profit.

It must be recollected, also, that the more the green crops are increased, the greater is the number of cattle you are enabled to feed, and, therefore, the more considerable is the stock of manure. It is very remarkable, and, however paradoxical it may appear, is nevertheless true; that, since the introduction of *assolements*, meadows diminish whilst cattle increase, and corn-fields diminish whilst grain increases. These miracles are performed by the artificial grasses, and the leguminous and other green crops, which not only prepare the earth for grain by their exudations, but enrich it by their remains; which leave no space for weeds, and supply abundant food for cattle.

CAROLINE.

And what is reckoned to be the due proportion of corn to meadow land in a farm?

MRS. B.

The distribution of a farm should be so arranged that the several portions should mutually contribute to each other's advantage. The farmer should aim at raising every year the same quantity of produce; for though it is true that the vicissitudes of seasons render this end unattainable, yet, by keeping it in view, you will the more nearly approximate to it.

When once the land is laid out to feed the number of cattle required for the work of the farm, and to produce the manure necessary for the soil, no change can take place without disadvantage. If you augment the produce of grain, it must be at the expense of the leguminous crops: the cattle will suffer for want of forage, and the soil from deficiency of manure.

EMILY.

And even the corn the following year will degenerate for want of that preparation of the soil produced by leguminous crops.

MRS. B.

The Belgians, whom we consider as among the best farmers, lay out their land so as to obtain, as far as possible, equal results annually. They derive their profit from the sale of their corn: this alone goes to market, the forage being all consumed by the farming cattle, and the manure employed on the soil. A Belgic farm consists generally from thirty to forty acres: the succession of crops is strictly regular, and comprehends a period extending from ten to fifteen years.

In rural economy an intervening crop is occasionally raised between two regular crops; thus, buck-wheat is often sown in that country as soon as the land can be ploughed after wheat, and is gathered in late in the autumn; but a double crop of grain is very exhausting to the soil, and it would be better that these stolen intermediate crops should be of leguminous plants. In England we do not attempt them: our corn is got in too late to admit of sowing for a second produce the same season.

EMILY.

What a prodigious advantage a rotation of crops has over fallows! If leguminous crops do not yield any profit, they defray at least all the expenses of their cultivation, and prepare the soil for a rich harvest of grain; whilst a fallow affords no crop whatever to repay the expense of ploughing and manuring, and does not so well prepare the soil for grain.

MRS. B.

The greater the variety of crops raised in a country, the more we consider that country as advanced in the knowledge of agriculture, for every new plant affords security against sterility; and the more crops are diversified, the smaller are the chances of suffering from the inclemencies of the season, for what is injurious to the one may be salubrious, or at least not detrimental, to the others. It affords also a surer market; for every species of produce will not fall in price at the same time, and thus the chances of loss are diminished. It is also an essential point, so to distribute the labour of a farm, that every operation may be made at the most suitable period.

EMILY.

The course of cropping must admit of modification, according to the locality, or the greater demand, for any one species of produce.

MRS. B.

Certainly; in England, for instance, where the beverage of the common people is beer, a greater quantity of barley and hops must be raised than in wine countries. Then the moisture of the English climate admits of our raising very abundant crops of turnips, peas, and beans: these plants enter with great advantage into our course of crops.

CAROLINE.

The vicinity of great towns must also influence the nature of the crops; it will be necessary to supply their markets not only with food but also with bulky produce, the carriage of which is expensive; such, for instance, as hay, a very great quantity of which is required to maintain the stock of horses and milch cows of a large town, which are quite independent of a farm.

MRS. B.

So far as regards their labouring for a farm, it is true; but the land profits by their manure: it is in order to supply hay for these animals that you generally see large towns surrounded by grass land. The oxen and sheep destined for food are brought from more distant parts, as they carry themselves to market almost free of expense.

The culture of the vine, especially in temperate climates, where this plant requires a great quantity of manure, necessarily modifies the *assolement*; for the farm must be so distributed as not only to afford manure for the succession of crops, but a large surplus for the vineyard.

EMILY.

This must be difficult to accomplish without making the general culture suffer from such a considerable subtraction; and, indeed, I have observed, that in Switzerland every thing seems to be sacrificed to the culture of the vineyard, as being that portion of the farm which affords the greatest profits.

MRS. B.

And which also occasionally produces the greatest losses. It may be considered as a game in which the highest stake is pledged; the greatest pains are therefore taken to increase the chances of winning.

The nature of the soil must also modify *assolements*. The light soils of Belgium and Alsace are very favourable to this system, while stiff tenacious soils offer considerable difficulties: they are, however, well worth the trouble of surmounting, as this mode of culture diminishes the quantity of manual labour, which such ungrateful soils require, and which renders their cultivation so expensive.

EMILY.

And what is the most eligible succession of crops?

MRS. B.

The most common is an *assolement* of only four years; the first of which is a hoed crop to destroy weeds: turnips, potatoes, beet-root, carrots, or any other plants with long roots, are very appropriate for this purpose, as it obliges the farmer to plough deep, in order to prepare the soil for them. After gathering in this crop, the leaves and remnants of the plants are ploughed into the soil, the land is manured, and wheat and clover are sown together.

The clover does not make its appearance till after the corn is reaped. Little advantage is made of the produce of clover the first season, but the following year it yields an abundant harvest. After having mowed it, the ground is ploughed, and the remains of the clover buried; and thus, both by its exudations, and by a part of its own substance, it renovates the soil after the exhaustion it had undergone in ripening the corn, and enables it to produce a second crop of grain the fourth year, which completes the *assolement*.

The rotation of crops must, however, necessarily vary with the soil: that which I have described, from M. De Candolle, is probably best adapted to France; in England turnips, I believe, are usually followed by barley, clover, and wheat.

CAROLINE.

Pray, why should not trees require an *assolement* as well as corn and leguminous plants? for the exudations of a tree during the number of years it lives must greatly injure the soil for another of the same kind. — Nay, I wonder how the same individual tree can thrive throughout a long life in a soil so deteriorated.

MRS. B.

You must consider, in the first place, that the roots of a young tree are of small extent, and both seek their food and give out their exudations in the ground immediately surrounding the stem. In proportion as they lengthen they extend their researches, spreading wider and piercing deeper into the soil; thus, after having exhausted it of nourishment, and deteriorated it near the stem, they find fresh aliment in a more enlarged sphere.

EMILY.

Then when a tree dies, if another of the same kind be planted in its place, the young roots will find the soil near the stem exhausted of nutritive particles, and vitiated by exudations? and yet, when a dead tree in an avenue is replaced by another of the same species it grows without difficulty.

MRS. B.

If you replaced it by one of another family, there is no doubt but that it would thrive better. One of the same species is, however, not without resources; for you must consider that the soil nearest the stem of the old dead tree has not been acted upon by the roots for a number of years; and, during this period of repose, it has been able, in a great measure, both to renovate its nutritive particles by the natural manure it receives annually from the fall of the leaf, and to purify itself from exudations of the old tree, these being absorbed by the grasses, underwood, and plants of various descriptions which surround the tree, and which, in return, supply the soil with exudations adapted to forest trees: these ameliorations will enable a second tree to live and grow upon the same spot as its progenitor, though certainly not with the same vigour as if it were of a different family.

CAROLINE.

In replacing a tree in an avenue we are not at liberty to choose its species; but in our gardens it is surely wrong to replace an old fruit-tree by another of the same species.

MRS. B.

This is not so easily obviated as you imagine; for it is not sufficient to change the species: you must change the family; and almost all our fruit-trees are of the same family. To remedy this inconvenience, the gardener must supply the young tree with fresh soil, which in a great measure answers the same purpose. This new earth should, if possible, be brought from the neighbourhood of forest-trees, which are of another family. Manure may at the same time be introduced. Nursery gardeners alternate plantations of fruit and of forest trees.

CAROLINE.

When a wood is cut down another springs up of the same trees, shooting up from the old roots, or germinating from the seeds naturally sown. Yet how can these young plants find sustenance in a soil both exhausted and vitiated by the parent-trees?

MRS. B.

You assume as a fact what is only a natural inference. If, when a natural forest be cut down, a second springs up, it will consist of trees of another family; a forest of oaks, for instance, may be succeeded by one of aspen, which is of a different family, or the oaks may be replaced by some species of fruit-trees.

CAROLINE.

I am not talking of replanting the forest, but of

that which would naturally spring up were the first cut down. Now we know that aspen and fruit-trees will not germinate from acorns.

MRS. B.

It is the acorns which will not germinate in a soil so ill prepared for them; whilst the seed of the aspen, or kernels of fruit-trees, which may chance to be here and there intermixed with the oaks, will find a soil so well adapted to them that they will germinate readily and grow rapidly. Thus a wood of aspen and fruit-trees will succeed to one of oaks; but, after a long course of years, the old stumps and roots of oak, favoured by the exudations of the forest of a different family, will shoot out, and, in the end, supplant the new forest, and a second forest of oaks will be re-established, but not till after an *assolement* of trees of another family. You see, therefore, that Nature occasionally makes successive as well as simultaneous *assolements.*

CAROLINE.

This is very curious. I did not conceive it possible for an *assolement* to take place where industry did not interfere with the natural course of vegetation. But this succession of crops must change once a century rather than once a year.

MRS. B.

They are doubtless of very long duration. The *assolements* of trees which occur in the course of agriculture are of a more transitory nature: they are generally made with a view to improve new soil, in order to prepare it for cultivation, such as the simultaneous *assolement* of broom and pines in the Campine of Belgium. These shrubs enrich the soil for the future cultivation of grain, both by their exudations and by the manure formed from their leaves.

I have seen a very singular *assolement* in the neighbourhood of the Rhine, consisting of alternations of vine and clover. After a period of twelve years the vines are grubbed up, and clover sown for three or four years.

But the most remarkable *assolement* is that of water. There are some districts in France in which the low grounds are laid under water for the period of a twelvemonth, and this is renewed every seven years.

CAROLINE.

What harvest can be obtained from such a culture, unless it be fish?

MRS. B.

Fish and wild fowl form, in fact, the only produce while the land is under water. This mode of culture has the advantage of draining the surrounding country, and of favouring the production of aquatic plants, which afford food for a prodigious quantity of worms and insects. All these productions, whether animal or vegetable, leave their relics at the bottom of the sheet of water;

and, when it is drawn off, the land remains covered with an abundant stock of the richest manure. There are many ponds of this description in the country of Bresse in the Lyonois. If the water with which these parts abound were more generally diffused, they would become marshy and unwholesome; for it is the scum of superficial stagnant waters which is deleterious, not the evaporation of deep waters. On the other hand, were these ponds permanent, their deposition of rich manure would either be lost, or could be drawn out only at a great expense; whilst, if you change the locality of these ponds, by drawing off the water to another spot, the manure remains ready spread on the soil, and the farmer has only to plough it in and sow his seed. The water in the mean time occupies another low land, where, in its turn, it accumulates and deposits its riches: with the assistance of locks, it is thus made to perambulate through the valleys and low lands. Great care is taken to preserve the young fish, and transfer them to their new basins; for these, like the sheep of the meadow, not only supply us with food, but enrich the soil for future vegetable produce.

I was once present at the operation of drawing off a sheet of water, of no less than seven hundred acres in extent. It was in the month of October. During the preceding summer, fish had been caught and wild fowl killed in prodigious abundance; but when the secrets of the prison-house were exposed to view, it afforded a very curious spectacle. The markets of all the neighbourhood were supplied with full-grown fish; the young fry were sold to stock other ponds; and rich and ample were the remnants of animal and vegetable manure which prepared the ground for culture the following season.

Simultaneous *assolements* are as advantageous to warm countries as successive *assolements* are in our northern climates. The circumstance chiefly to be attended to in this mode of culture is, that the two crops should seek their nourishment at different depths; thus the vine and corn may be raised together, the roots of the vine being much longer than those of corn.

EMILY.

In Italy we have seen them continually accompanying each other: strips of corn separating the rows of vines trained on trees; which latter also compose a part of the *assolement.*

MRS. B.

The vine is sometimes twined round the olive, whose roots strike still deeper into the earth. For the same reason, the peach and the almond are often raised in vineyards; while apple and pear trees would not thrive, because their roots are as superficial and as spreading as those of the vine.

CAROLINE.

Their shade might also be prejudicial, while the foliage of the peach and the almond is comparatively light.

MRS. B.

The degree of shade must be regulated by that of the temperature of the climate. In hot countries leguminous plants succeed well interspersed with trees, because their shade, by diminishing evaporation, retains moisture in the soil. Thus corn thrives intermixed with turnips and clover: the two latter, when young, requiring the shade which the corn affords; and after it is reaped, the sun is necessary to ripen them.

Wheat and rye are sometimes sown promiscuously: it is an old custom, and, I believe, a very bad one; both plants being of the same family, their exudations are noxious instead of advantageous to each other. Then they do not ripen at the same period; so that when reaped, the one must be over-ripe, or the other not come to maturity. If the intent be to make bread of these two species of corn, it would be preferable to mix the grain after the harvest; indeed, it would be best to keep them separate till after grinding, for, not being of equal size and hardness, a loss is also experienced in grinding them together.

CAROLINE.

It is customary, also, in sowing grasses for forage to mix a great variety together.

MRS. B.

It is supposed that the species which is best adapted to the soil will thrive so well as to choke the others; but it would be a more judicious mode of proceeding, to try by experiment, which kind of grass was best suited to the soil, and sow that alone.

On the confines of the cultivation of vineyards, that is to say, in those latitudes where the vine with difficulty ripens, the cultivator aims at producing a large quantity rather than a superior quality of wine. For this purpose the vines are frequently trained on trees, which multiplies the fruit at the expense of its flavour.

CAROLINE.

It is singular that the same mode should be resorted to, in climates which are too cold, as well as in those which are too hot for the vine. In Italy they are trained on trees, to afford them shade; but on the cold confines, shade must be very prejudicial, more so, I should have thought, than would be compensated by the increase of production.

MRS. B.

On the limits of the vine countries, the great demand for common wine, in order to avoid the expense of its carriage from more distant parts, ensures a sale for wines of the lowest description.

Maize or Indian corn forms an *assolement* with peas and French beans: it affords a support to these climbing plants; and, being of the grass tribe, its exudations are favourable to leguminous plants.

EMILY.

It is inconceivable what an abundance of produce the earth yields under the influence of a southern sun. In Tuscany we have seen flourishing together, in the most perfect harmony of culture, the olive, the vine, corn, and a variety of leguminous plants.

MRS. B.

And yet the soil of Tuscany is not very favourable to vegetation. It is, indeed, well cultivated, the Tuscans, after the Belgians, being esteemed among the best of agriculturists; and they have, as you observe, the advantage of a most prolific sun. It is for this reason that they, in common with the cultivators in warm climates, aim at producing a numerous simultaneous *assolement;* whilst the Belgians, with the inhabitants of other temperate climes, must content themselves with a succession of crops. The *assolements* of a Belgic farm, we have observed, extend from ten to fifteen years, the farm consisting generally of about forty acres; in Tuscany they are usually circumscribed to fourteen acres, with a soil inferior to that of Belgium; yet the more ardent sun of Italy produces a result nearly similar. In Tuscany the farmer is not obliged to rear his crops in slow succession; they are poured upon him, as it were, from the cornucopia of abundance: oranges, lemons, olives, melons, peaches, corn, and vegetables spring up together, to delight his eyes and to heap his board.

It is remarkable that these two countries, now so distinguished for agriculture, were once no less celebrated for their commerce.

EMILY.

This seems reversing the natural order of things; for, in general, it is the abundance of agricultural produce which leads to the establishment of manufactures and trade.

MRS. B.

That is certainly the most usual mode of progressive improvement. On the other hand, when a people enrich themselves by commerce, it is a very natural consequence that they should lay out some of their wealth in the improvement of land. Then it so happened, that, as Europe advanced in arts and civilisation, commerce, which began first to flourish in Tuscany and Belgium, and was, indeed, almost exclusively confined to those countries, became more generally diffused. Political events also tended to diminish the trade of these countries; and, when it fell into decay, agriculture proved a fortunate resource for the wealth and industry of the people. They transferred to this employment not only their capital but that spirit of speculative enterprise, wisely regulated by those habits of calculation and of order, which distinguished them as merchants; and, when engaged in any hazardous experiments, the regularity of their accounts gave them exact results, and showed them whether they ought to be prosecuted or abandoned. This union of energetic vigour and methodical arrangement has achieved the wonderful enterprises of these excellent agriculturists.

END OF THE FIRST VOLUME.

LONDON:
Printed by A. & R. Spottiswoode,
New-Street-Square.

Printed in the United States
By Bookmasters